ADVANCE PRAISE FOR

electric *water*

With *Electric Water*, Christopher Swan provides a fresh view of how present-day technology will enhance the human condition. This is a must-read for everyone with a basic knowledge of technology. I am stimulated by this easy-reading, amazing, and enjoyable book.

— PAUL G. HEWITT, author, *Conceptual Physics, 10th Edition*

Electric Water is a timely read — the fossil fuel era is finished! Christopher Swan's visions rank with those of Amory Lovins, Al Gore and a few others who know that the coming solar era will re-energize humanity in a harmonious, sustainable, more democratic way. It's high time we learn to operate Spaceship Earth using renewable energy and resources available.

— DAVID WRIGHT, AIA, Environmental Architect

Electric Water is a healthy dose of common sense! In an increasingly confused world with an obvious lack of vision, critical thinking and ethical leadership, Christopher Swan delivers a manual for concerned citizens for better stewardship of their Mother Earth. This book should be required reading for anyone joining in support or teaching in any school or environmental organization.

— BRUCE L. ERICKSON, international management consultant
    for Creating Sustainable Healthy Communities

electric *water*

THE EMERGING REVOLUTION IN WATER AND ENERGY

CHRISTOPHER C. SWAN

NEW SOCIETY PUBLISHERS

CATALOGING IN PUBLICATION DATA:
A catalog record for this publication is available from the National Library of Canada.

Cover design/digital composite by Diane McIntosh.

Printed in Canada.
First printing August 2007.

Paperback ISBN: 978-0-86571-585-1

Inquiries regarding requests to reprint all or part of *Electric Water*
should be addressed to New Society Publishers at the address below.

To order directly from the publishers,
please call toll-free (North America) 1-800-567-6772,
or order online at www.newsociety.com

Any other inquiries can be directed by mail to:

New Society Publishers
P.O. Box 189, Gabriola Island, BC V0R 1X0, Canada
(250) 247-9737

New Society Publishers' mission is to publish books that contribute in fundamental ways to building an ecologically sustainable and just society, and to do so with the least possible impact on the environment, in a manner that models this vision. We are committed to doing this not just through education, but through action. This book is one step toward ending global deforestation and climate change. It is printed on acid-free paper that is **100% post-consumer recycled** (100% old growth forest-free), processed chlorine free, and printed with vegetable-based, low-VOC inks, with covers produced using Forest Stewardship Council-certified stock. Additionally, New Society purchases carbon offsets annually, based on an annual audit, operating with a carbon-neutral footprint. For further information, or to browse our full list of books and purchase securely, visit our website at: www.newsociety.com

NEW SOCIETY PUBLISHERS   www.newsociety.com

To Sandra

for the gift of our sons

and a vision of life.

Contents

Acknowledgments

THANKS TO Maggie Swan for hearing me out, Robert Swan for sharing the world and my sons Kieran and Torrey for their inspiration. A big thanks to Raymond Dasmann and Gary Snyder for sharing their visions many years ago. Applause for Caitlin Collentine for her encouragement. Thanks to Louise Lacey for asking me to share the vision; to Jeff Lohrmann and Gene Harmon for reading and commenting; and to Jilla Levington for moral support. A big thanks to Dan Osborne, Michelle Watson and David Vasquez for assisting in the envisioning of technological realities. A tip of the hat to New Society Publishers, and especially Christopher and Judith Plant for their faith, and Murray Reiss and Ingrid Witvoet for their patience. A bow to David Kupfer for drawing me out of my shell, and to Bruce Erickson for asking me to write about water. A toast to Chet Roaman, for being the best of friends and the finest of teachers. And a prayer for the golden eagle who landed on my windowsill.

Introduction

MANY OF THE VISIONS sketched in these pages began to emerge four decades ago, when I was a young man in San Francisco in the sixties. I came of age with the support of a father who often took me along on his trips as a salesman marketing industrial products. By the time I was 20 I'd ridden in a caboose behind a steam locomotive, toured nuclear submarines and aircraft carriers, as well as various uranium mines, lumber mills, steel mills, power plants and oil refineries, and learned about cars by rebuilding one.

In the 1970s my fascination with architecture and infrastructure melded with a growing concern for the environment, and evolved into a vision of new transport, energy and water technologies, and a new approach to the land. This vision emerged in a science fiction book, *YV 88,* which outlined how a solar-powered railway operating on landscaped track could allow the ecological restoration of Yosemite Valley.

In the 1980s I formed a company to develop new railways. In 1991 we proposed the Yosemite project to the community and I found myself in the curious position of having written a science fiction book only to later be the developer of that vision. With the support of the park superintendent, executives of the park's concessionaire and environmental groups, I outlined the program before Congress and began to seek financing, but this proved too elusive and we had to shelve the project.

By this work I came to recognize the depth of our environmental plight and began to consider ecological restoration. But restoration was not even being discussed at the time; most environmentalists believed all that was gone was gone forever. I didn't agree. I once asked the eminent oceanographer Jacques Cousteau if, in all his travels, he'd seen any evidence of a civilization that had recognized it was destroying its ecological foundation and then restored it. He said no, but then added, with a wry smile, that if it was ever to happen it would happen now.

That was 28 years ago and the turning point occurred not long after that conversation. The first restoration projects began to appear in the news. Initially they were tiny efforts, a pond or a stretch of a creek, but soon the scope expanded dramatically. The significance of restoration is underscored by Jared Diamond who, in his recent book *Collapse: How Societies Choose to Fail or Succeed,* cites ample evidence for how environmental changes effectively set the stage for the economic collapse of one civilization after another. Today restoration is a global effort, often involving major regional initiatives measuring in the billions of dollars. This has never happened before.

There were other turning points in the early eighties. Railroads awoke from a long sleep, in the form of a new trolley line in San Diego and growth in rail freight traffic, and the first 100 percent solar-powered homes began to appear in the backwoods of the US. These trends would prove to have profound implications.

All three trends, as well as parallel events in other fields such as organic farming and electric cars and high efficiency appliances, confirmed my belief that we could have ample energy and a high quality of life without continuing to transform the planet into a machine. I knew there need be no link between our quality of life and the destruction of earth, as if our seeming success and ecological health were mutually exclusive.

We are now entering a period when two major issues are likely to dominate our attention for some time: the decline of oil and the rise in the planet's temperature. If we do nothing to alter our course these two trends will, to varying degrees in various places, contribute to conflicts, starvation, disease and economic collapse. The causes behind these issues are not a result of normal events in the course of civilization's evolution, but of our use of extraordinary and unsustainable technologies over the last two centuries, and our inability to fully grasp the unprecedented magnitude of our numbers and our power.

We must recognize the uniqueness of our time. Modern civilization, beginning with the railway in the 1850s, is wholly unprecedented. While we may see similarities in human behavior and politics with long-gone civilizations, there is little similarity in our capabilities, achievements and potentials.

Today I see well-established trends pointing to a world where cities not only rely on the sun, but where the wild may be as close as the nearest shoreline, forest or grassland; where everyone would have equal access to mobility, healthful food, ample energy and pure water, and no one's life, nor that of the earth, would be compromised by anyone's use of any resources.

I do not see a bleak future. I've been accused of being a Pollyanna because I don't share the prevailing and often cynical visions of our future. I do see many potential horrors that *could* occur, and many that are likely to occur. But outside Hollywood there's no future in horror.

I turned away from such notions to focus on the visions I saw, reinforced by the ideals of millions of people all over the world who shared my feeling that modern life had to be transformed, not just fixed. These individuals now represent groups numbering in the millions, and they are focused on all manner of environmental, cultural and technological concerns. This spontaneous arising of citizens as activists and entrepreneurs has been a source of faith, sustaining me through many lean years.

I've witnessed this grassroots movement over four decades, following countless innovations in land restoration, energy, water, transportation and several other fields. I've also witnessed a denial of these realities by mass media and major institutions. To this day I find many popular debates persist as if solar power were not the fastest growing source of energy in the world, as if railways were not being expanded all over the world, and as if ecological restoration, organic farming and a raft of other trends were mere fantasies. Intelligent people have told me these things were not happening because the media or the president would have told them. They do not see a revolution happening all around them because it isn't being televised.

The vision I've been given has been sustained by faith in people, and their faith in me, and it has been informed by knowledge of basic infrastructure. I have long been a student of infrastructure, not just the technology of transport, energy or water, but the culture, economics and experience of it all. Major changes in energy, water or transportation—changes that will leave existing energy sources behind and may result in ameliorating if not reversing global warming—are tangible possibilities involving real technologies being made right now. History reveals that we've changed infrastructure much more rapidly than most people think, and precedent suggests all the ingredients are now present for a sweeping transformation.

Electric Water is about visions. It is not a book of prescriptions or analysis, but a journey into the realms of technology, culture, economy and our future. It is my hope that I might share with you what I believe is a most profound vision of a new world now emerging.

PART ONE

Where We Are

1

In Light of Water

On New Waters

WE ARE ONE. We are walking columns of water. We are striding statements of chemistry and geology, bearing with us the legacy of where we lived and what we drank. We are each a structure carrying around a brain made largely of water, yet containing some vague notion called "will," a force that causes us to shoot electrons around our interior sea, call the buzz an idea, then vibrate some air in the direction of a fellow human's ear, and bounce the idea off them.

Ideas are passed on like genetic codes of intellect. Ancient civilizations leave us with artifacts exhibiting a unity of vision, where the daily lives of all people were linked to the spiritual values of a civilization and expressed in magnificent displays of pictographs, statues and innumerable works of monumental architecture. These unique visions arose in fertile valleys to form the whole cloth of a civilization sharing one vision, only to fade as the weave of reality came apart, its threads of language, history and culture worn bare by wars, disease, poverty or merely the ideas of another civilization. Yet while a civilization's accomplishments may be forgotten a wisp of vision may remain, just as one might forget an alluring face, but not the sensation of seeing beauty. The Incan civilization is gone, yet the ruin of Machu Picchu teaches us.

The work of ancient civilizations speaks to us because they were faced with the same realities we face today. Over thousands of years people have organized themselves into ever more complex civilizations. Over all that time billions of people have rediscovered the basics of human life; of elemental issues of earth, air, fire and water; temporal realities of shelter, food, health and security; and spiritual realities of birth, life, death and beyond. These realities have been woven into stories in memory, stone, paint, ink and pixel.

Yet such basic realities can be hard to perceive in modern society. Contemporary "reality" is sliced into specialties, diced into jobs and sautéed in the oil of self-interest. Modern technology demands a high degree of specialization, but the price is a society divided into unprecedented categories and further subdivided into a ranking of jobs. Millions of people relate to a professional community of like-minded folks more than a physical community of many-minded folks. Specialization is compounded by the normal tendency of all religions, governments, corporations, committees, clubs, families, tribes and couples to want to believe their way is best and all other religions, governments, corporations, committees, clubs, families, tribes and couples are idiots. On second thought they're nice people, just not of *our* caliber.

The result is a society of tribes lacking a common vision, or a common understanding of what the commons even is. The utterly normal tendency of everyone to want to differentiate themselves at the local watering hole has been heightened by media marketing to where even the slightest differences are exaggerated and commonality ignored. The simple and common fact of earth, air, fire and water as primal elements of life we all share is treated as if it were arcane wisdom in ancient texts.

Industrialized societies are at war with themselves, not simply because of normal political conflicts, but because they lack a common will around a common purpose. In the US, religious media empires make it possible for the faithful to exist in the United States of Singular Religion and never venture outside, while gated communities for wealthy people are designed so the real world can't intrude. Whose "real" world? What's real?

Nevertheless there is a yearning for a common purpose. On September 11, 2001, the city of New York and the world shared a tragedy in the attacks on the World Trade Center. On countless occasions since the arrival of global instantaneous telecommunications the world has come together around major events, from earthquakes to hurricanes to a massive tsunami to the death of Princess Diana. We, as residents of the world, feel empathy and express it in a common purpose of assistance or grief.

There was a time not long ago when the common realities all people shared were of great importance, and were a subject of continual specula-

The commons has no voice in US society beyond the words and pictures of various special interest groups. Mass media is almost entirely devoted to personal and political stories, and almost entirely sponsored by companies selling personal products. After decades of this bias, as exemplified in televison commercials, the majority of Americans have become so focused on personal concerns and personal products they no longer understand the infrastructure upon which their lives, and all those products, rely.

tion and discussion around everyone's table. We shared the realities of farm and town, of crops and weather, of markets, factories and local employment, and of the health and the life of our community. Now hundreds of millions of people live all over the map, detached from any relationship to a physical place. They live in virtual worlds where personal concerns and community concerns may be defined in a myriad of ways; where one might ask, "Do you mean my physical community where I live or the global professional community I relate to?" Yet at the same moment in history when millions think nothing of spanning an ocean over dinner and drinks, when millions define their lives around the most arcane of interests and may barely know their next-door neighbors, most people know they are primarily made of water and rely on air, both of which they share with everyone else. Millions of people might also know that we humans, along with all animals and plants, constitute part of a vast living entity carpeting the land and permeating the seas, defined by earth, air, fire and water.

We know we share this place called Earth. We know that the earth we stand on, the air we breathe, the fire we start and the water we drink are all of a piece. We can know that the global problems we now face are also of a piece. We are faced with nothing less than a need to transform how we relate to earth, air, fire and water.

A Common Reality of the Commons

The commons is not just a plaza at the mall. It is all that we share in every breath; it is all the lakes, rivers and seas, and it is all the world of all the plants and animals who exist by our side on this blue ball in black space. The commons is the water, energy, transportation and communications systems we share, the health care, judicial, police and fire services we support, all the parks, museums, stadiums and schools we sustain, and all the laws related to international trade. It is the wind we feel but cannot control; the world of institutions we are born into, but did not choose; the long waves of generations that came before us, leaving a legacy that presses us to repeat our past in the name of tradition and our "common" ancestry. The commons is everything we do not define, control or buy solely at our personal or family discretion. It is everything not for sale, not commodified and therefore usually not on television.

The commons is also the story of how we deal with the commons. It's the body of laws, policies and unwritten understandings that relate to all that we share. It's simple understandings, such as "we don't swim in your toilet so please don't pee in our pool," a sort of southern California golden rule, and it's a veritable catalog of laws about everything from local land use to the shade of the yellow used down the middle of the road.

Advocating for anything related to the commons is like trying to sell air. Everyone understands cars or appliances or houses, or common political issues like education or crime, but few understand county general plans or electrical grids or complex infrastructure programs. Virtually everyone has seen millions of TV programs sponsored by personal products, and only a tiny fraction the programs about public life, sponsored by all of us. Public broadcasting, as well as several commercial channels, offers frequent documentaries on nature, political realities, history and new trends in technology, as well as sophisticated dramas about gritty public issues. For every second of such programming there are innumerable minutes emphasizing how one's personal life could be improved by consuming more of the personal products depicted in the commercials. There are advertisements for bottled water, not rivers.

We see the car, not the highway; the water, not the pipes; and the wall outlet, not the coal power plant 200 miles away. We see what we own, not what we share. Yet we share a commons of infrastructure. The architecture of our homes, farms, factories and office buildings may be different, and we may choose different brands of phones, refrigerators, stoves, toilets, cars, bikes and all manner of other products, but we share fiber-optic cables and satellites when we communicate; highways, airways and railways when we travel; huge ships when we buy products; reservoirs and aqueducts when we drink or wash; power plants and trillions of miles of wire when we turn something on. We even share do-it-yourself literature covering all aspects of fixing, financing or building all the products we use. But there is no do-it-yourself book for the infrastructure we share because one person can't do it alone.

We share infrastructure without regard for any one city or state, nor any one religion, culture or race. Brits can take a high-speed train from London to the Swiss Alps; North Americans think nothing of driving across the continent; and a very large portion of the world's population doesn't hesitate to hop on a plane to a distant city with a different culture and religion. Virtually everyone calls around the planet as if they were just yelling at a friend across a crowded room: "Hello?"

We share a profound incentive to ensure that our infrastructure works smoothly. Common infrastructure is inviolate. In many cultures damaging infrastructure is a serious offense. Destroying crop land would be a capital crime. Polluting the air could make one hated. Extinguishing the common fire would be an act of hostility. Openly polluting a well could get one summarily killed. Earth, air, fire and water.

Today we may not share a common language of the four elemental forces, but we share a common experience. We all start as cells multiplying in a tiny drop of water. We start inside mom and become what we become in large part because of the water she drinks and the food she eats. This water we take in, from within the womb and beyond, is first a trickle, then a tiny river that grows with us as we become part of this assemblage of walking water called humanity. Its dynamic tension, its capillary attraction and its consistency makes water the substance that glues us together. We are all in that river.

Given what we share, from our fundamental biology to our common history, from the promise of technology to the ongoing 24-hour a day operation of a vast infrastructure stretched across nearly the entire planet, it would seem we have built a common reality that transcends any cultural or political division ever invented. We do not necessarily share languages or religions or culture, but we do share a language about all the technology we rely upon. From railways to jet planes, and from dental instruments to machine tools, the world shares thousands of common terms for all sorts of machinery, as well as common graphics for road signs and emergency symbols. There is only one language spoken by international airline pilots worldwide, — English.

Despite its mundane origins this infrastructure we share is far more profound in its impacts on our lives than any religion. This commons represents the potential to magnify our capability — our power as a species to destroy or enhance the commons we share.

In the sixties the confluence of environmental awareness, the space program and computerization, plus a dose of radical politics, spawned a new vision of the commons. Originally confined to the US, Canada and Europe, it has since migrated to other places, as well as informed industry and government policy. It is now global.

What distinguishes this new vision of the commons from the old commons of 20th Century industrial society? This new vision is a loose collection of ideas and attitudes evident in the personal beliefs of millions of people and the policies of countless organizations. It reflects subtle cultural shifts from emphasis on lone competitors to partners, from complex hierarchies driven by external authority to small groups managing on their own; from information by central agency to information available to all; from seeing life as a mechanical system of parts to seeing life as one vast interrelated ecosystem; and from viewing resources as scarce to recognizing an abundance if we use what we need with efficiency born of reverence.

This new vision of the commons embraces new technologies in the realms of water, energy, transportation and land use. But unlike popular utopian visions of the past, invariably dominated by technology, this new vision embraces the wild. It includes a trend of extraordinary significance: ecological restoration.

A confluence of trends

Due to the propensity of people to focus on the negative, or the inaccurate, and the tendency of mass media to focus on the conflicts and horrors of life, there is a lack of awareness of trends that have been evolving for decades. These trends represent an alternative pathway now firmly established, growing rapidly, and largely initiated long before climate change was even an issue. Some trends, historic preservation for instance, are valuable by what they do not do, such as consume more energy to build a new structure.

There is ample evidence that innumerable ancient civilizations declined not because they lost a war or a leader, but because their numbers exceeded the capacity of the land and water to support them. The farmers consumed the soil, herders' animals denuded the grasslands, hunters killed all the game, fisherman caught all the fish and they all ate their own economy.

There is no apparent historical evidence that any civilization consciously recognized the ecological source of its decline and began restoration. People simply moved on. But then we reached the late 20th Century, the 1960s, when billions of people concentrated around the world's fertile valleys were busy accelerating the decline of ecosystems on a global scale. It thus became obvious to small but influential groups of scientists, artists and citizen activists in the US and several other nations that a global environmental catastrophe was inevitable if we, as a species, continued not to heed the signs.

In 1980 there was no ecological restoration movement. The subject was just an idle dream of a few environmentalists. Then projects began to spontaneously emerge, in several places. Although environmental impact legislation in the US did trigger some major restoration projects, such as coal mining companies restoring Wyoming grasslands, there was no government edict, no major program nor even a dramatic book or movie. It just happened. Today there are hundreds of ecological restoration projects all over the world, and several projects have already resulted in plants and animals returning after a long absence. This has never happened before.

We are entering new waters. Yet we are also entering ancient waters, as though subconsciously recognizing the knowledge in the counsels of ancient civilizations and the implicate wisdom in the flow of all waters. We call it a blue planet because it is permeated and nearly covered by water.

The Wolf We Know

No doubt the conversation has been replayed a million times in millions of households and at social events. It's the state of the Earth conversation. It centers around the assertion that ecological problems will do us in if we do not change because we cannot sustain ourselves by destroying the very ecosystem we rely upon.

We could assume the proponents of doom hate people. We could assume ending reliance on oil will be taken care of solely by market forces. We could assume global warming is not that bad. We could assume our flooding the waters with inorganic chemicals is unrelated to the lack of fish in the sea.

We can argue about when we will run out of oil, or how rapidly China and India will consume oil at current exponential growth rates, but we cannot argue with the end of oil. We can argue about the specifics of warming or cooling, but we need not argue that ice is melting faster and freshwater from *above* sea level is cascading into the sea. We can argue about the quantities of carbon dioxide or water vapor or nitrogen, and how fast the excess of these gases will transform the world's ecosystems, but we cannot argue with melting permafrost and rising methane that will kill virtually all life on Earth if temperatures keep rising.

We could consider that climate change and peak oil are not guesstimates of our future, but estimates of our present situation, and the key problems we must address if civilization is to survive, and life on earth is to survive. Fossil fuel use leaves behind greenhouse gases that are changing the climate and the composition of the atmosphere. It also produces substances that are changing our bodies, and pollutants that are changing the acidity of oceans and the health of whole forests and threatening the very existence of vast ecosystems.

In the controversy over the extent and implications of climate change many people will deny the scope of the problem, as if assuming humanity could not possibly have such massive impacts. But few people alive today are aware of just how big cities have become. A century ago a big city was 500,000 souls; now many big cities are 10 to 50 times that size and growing. This unprecedented scale is matched by prodigious resource consumption and the inevitable production of dust, soot, heat and greenhouse gases, all of which affect respiratory health, ecosystem health and global weather.

It may be tempting to blame the messenger, or the government, or the corporate world, or some religion or philosophy. But if we're to address the problems we face, we must acknowledge we are beyond the morality of the play because we are the play. We are *all* confronted with these problems because they concern the very air we share. Earth doesn't care whom we blame.

In the debate over global warming there are three major opinions. One, the science is bunk. Two, the science may or may not be bunk, but it's happening and it's either an act of God or nature depending on whether one is "religious" or "spiritual." Three, the science is solid, and clearly reveals the link between our industrialization and changing climate — we did it.

Earth is a large sphere enveloped by a film we call the atmosphere. This thin soup of oxygen, hydrogen, carbon dioxide, nitrogen, methane, carbon monoxide, ozone and several other gases, plus various chemicals synthesized by animals and plants, plus new chemicals we have introduced, plus lots of dust, is roughly 60 miles thick — an hour's drive. This wall of gas also acts as a blanket retaining the sun's heat, as well as the heat produced by anything burning on the planet.

Global climate change is caused by gases and water vapor from the burning of forests and fossil fuels, use of petrochemical fertilizers, and the heat off buildings and pavement. The gases normally act much like a translucent mirror reflecting some of the sunlight bouncing off the seas and land rather than letting it all escape into space. Add too much carbon dioxide, nitrogen, methane and water vapor and the temperature rises. Now the temperature of the atmosphere is causing permafrost to melt, exposing long frozen plant matter which begins to decay, causing methane to rise and add to the warming. Too much of these gases in the atmosphere increases its function as a

Given the bleak realities before us it all might seem hopeless. In this context it's important to recognize how quickly infrastructure has been developed — therefore how rapidly we could reverse the situation —– and how many of the technologies we can employ represent work dating back decades. Ecological restoration began with the work of Aldo Leopold in the 1930s. Practical photovoltaics were conceived in the 1950s. Electric cars have been in development for over 20 years.

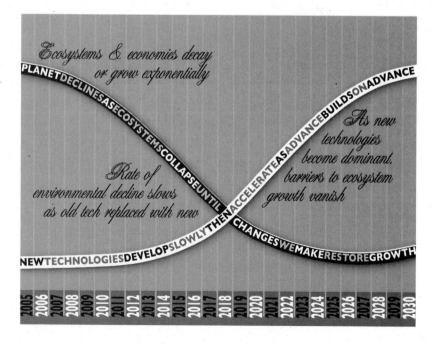

heat mirror, the "greenhouse effect," in turn causing temperatures to rise still further. Methane is poisonous.

The science is stark. Current research clearly suggests that as temperature rises the process builds exponentially, causing ice caps atop land to melt and sea levels to rise at an accelerating rate. This raises the risk of farmland and urban areas becoming permanently flooded. Cities like New York might be able to hold back the waters with dikes for a decade or two. The sheer area of coastal lands would preclude such strategies in most communities. Perhaps in 50 or 100 years, New York City would either be an island or the new Venice, with valet mooring for boats and flotillas of yellow water taxis cruising down Seventh Avenue. Climate change is viewed by some US Department of Defense planners as the defining national security issue of the next *several* decades. Terrorism would be a minor sideshow.

If we deny the science outright and assume it's a natural event or act of god this effectively translates to no action because there is nothing we can do. It is out of our hands and we're off the hook. We're also dead. Rising atmospheric and ocean temperatures will unleash the earth's reservoir of frozen methane and all life on Earth will be over.

On the other hand, if we accept the correlation between industrial growth and greenhouse gases and the evidence oil is running out fast, then we are faced not with a disaster of unknown origins, where efforts to address the problem would be just wild guesses, but a problem of understandable origins, conceivable solutions and somewhat predictable outcomes. Given that the 20th Century was doubtless the most recorded, studied, analyzed and chronicled century in all history it's fairly easy to determine what happened.

Accepting the magnitude of our situation demands we also accept the imperative of extraordinary change on an unprecedented scale. Over the last several decades millions of people, worldwide, have accepted this challenge. They comprise a vital movement of environmentalists, entrepreneurs and community activists, as well as scientists, engineers, architects and other professionals, all working in nonprofits, companies and government agencies developing new green technologies.

Among those who have long advocated sustainable strategies it is widely accepted that the systems of modern industrial society were built around unsustainable assumptions. Nature's wealth is not inexhaustible, price is not the only factor deciding what species, resources and landscapes are *used* and consumption is no one's key to happiness. Sustainability advocates recognized how endless resource consumption to feed endless population growth leads to endless modification of ecosystems and the end of civilization. They recognized the absurdity of a business plan that destroys its own market.

What they developed over the last 40 years is a recipe for the transformation of the planet. It is emerging in the work of major corporations, especially in fuel-cells, electric vehicles, water purification and photovoltaic cells. It is visible in growing projects to restore rivers, grasslands and forests, as well as the quiet return of wild animals to the city. It is obvious in the rapid rise of organic farms, natural food stores and wild game ranching. It is expressed in thousands of new green building products, and buildings that use less and less energy. It is evident in the growth of research in renewable energy, climate science and ecology. It is experienced in a thousand malls and big cities where new environments make walking preferable and riding a bike pleasurable.

All these trends point to a new world growing in our own backyards. This new vision didn't happen by fiat, nor by corporate mission; rather it grew spontaneously all over the world. Moreover, it is not confined to technological change just for efficiency or profits, but seeks technologies that support positive social change. Organic farming is not just a matter of healthful food, but of reduced health care costs. Renewable energy is not just clean environmentally, it's accessible to all people and far less vulnerable to terrorism and natural catastrophes. All the new technologies and related strategies represent not just a potential solution to global warming and the end of oil, but a considerable improvement in quality of life.

In 2005 and '06 the world's capital markets recognized the significance of climate change and the imperative of green technologies. Unlike biotech, which has absorbed huge quantities of capital over more than a decade with little revenue to show for it, most green tech involves low-tech strategies, no-tech retailing or high-tech manufacturing. No big breakthroughs are required, and there's a ready market that's large and global. Capital, by the billions, is flowing into green tech. Dollars are already known as greenbacks.

But the wolf hasn't left the building yet.

We Have No Time

Those who study the geophysical data, who examine layers of glacial ice to gauge climate over millennia or seek patterns in the statistical summaries of weather over centuries, adhere to the scientific protocols of assessing data without drawing premature conclusions. As a group, scientists do not generally make definitive statements about what will or won't happen in the future. However, some scientists and thoughtful journalists familiar with the data have speculated that the quantity of greenhouse gases already in the atmosphere, plus rising heat from cities and the cultural barriers to stopping these processes, suggests that even if we ended fossil fuel use tomorrow the process we've inadvertently set in motion seals our fate. They argue that it's already too late.

Other scientists and professionals who report on science acknowledge that implicit in the forecasts of dramatic ecological changes made by the Intergovernmental Panel on Climate Change's studies (released in the spring of 2007) is the assumption humanity will have no choice but to adapt. They resign themselves to a world where countless numbers of people migrate to higher ground and accept the loss of billions of acres of farmland and once-productive coastal shallows that took millions of years to create; and where millions of us "adjust" to further degeneration of forests and grasslands, while accepting the loss of millions of people to starvation and civil chaos wrought by ecological collapse. Such a grim scenario of acceptance may be "realistic" in the realm of rational scientific prognostication, but it ignores the staggering emotional dimensions of such losses. No doubt the species could adapt, but the price to civilization would be catastrophic. This scenario also ignores the risk of rising heat. If temperatures rise and stabilize at a slightly higher level we might muddle through, but if average temperatures of air and sea continue to rise, the portion of methane released into the atmosphere will increase until life on the planet is diminished, if not extinguished.

Such a grim scenario is "realistic" as an assessment of the disaster we face, but it ignores how every problem spawns a solution. Countless environmental studies, done for all manner of building projects and scientific objectives over the last several decades, were studies of what is, not what could be. They commented upon events that happened without acknowledging how their comments would affect events yet to happen. Remarkably, there's little popular recognition of how 40 years of environmental studies, and the grim pictures they usually painted, have long motivated millions of people to develop innovations in all manner of technology in order to mitigate or eliminate an environmental problem.

As a species we have never faced such massive problems on a global scale, but as nations we have united and accomplished extraordinary achievements. Following the Japanese attack on Pearl Harbor in December of 1941 the US unleashed a staggering torrent of manufacture. Companies changed from making cars to making tanks, from making steel to making whole ships and from making pots to making helmets. Weapons manufacture stepped up from a dozen a week to a dozen a day to a dozen an hour — of everything. By the summer of 1942 any German or Japanese officer who knew the details of the productive capacity arrayed against them knew the war was over, but for the naive illusions of the Axis fascists. We could and would out-produce them. End of war.

Following World War II the United States managed the Marshall Plan, a four-year, $13 billion program to rebuild Europe — $140 billion in 2006

Parallels exist between the first 20 years of the 20th Century and today. Many of the same issues are with us again. Media and retailing have just gone through a transformation due to computers and the Web — then it was the Sears catalog, now it's the Amazon Website. After a century of absence electric cars and railways are back on the agenda, as are new energy systems. Remarkably, many of the inventions that defined the 20th Century — cameras, subways, cars, planes, movies, recordings and skyscrapers — all became common knowledge within only a few years.

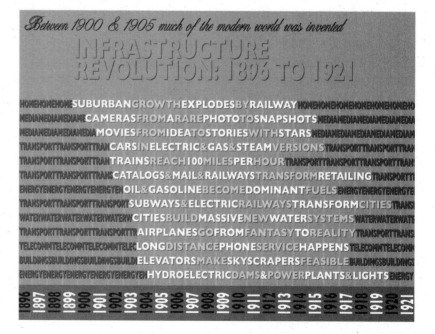

Between 1900 & 1905 much of the modern world was invented

INFRASTRUCTURE REVOLUTION: 1896 TO 1921

SUBURBAN GROWTH EXPLODES BY RAILWAY
CAMERAS FROM A RARE PHOTO TO SNAPSHOTS
MOVIES FROM IDEA TO STORIES WITH STARS
CARS IN ELECTRIC & GAS & STEAM VERSIONS
TRAINS REACH 100 MILES PER HOUR
CATALOGS & MAIL & RAILWAYS TRANSFORM RETAILING
OIL & GASOLINE BECOME DOMINANT FUELS
SUBWAYS & ELECTRIC RAILWAYS TRANSFORM CITIES
CITIES BUILD MASSIVE NEW WATER SYSTEMS
AIRPLANES GO FROM FANTASY TO REALITY
LONG DISTANCE PHONE SERVICE HAPPENS
ELEVATORS MAKE SKYSCRAPERS FEASIBLE
HYDROELECTRIC DAMS & POWER PLANTS & LIGHTS

1896 1897 1898 1899 1900 1901 1902 1903 1904 1905 1906 1907 1908 1909 1910 1911 1912 1913 1914 1915 1916 1917 1918 1919 1920 1921

dollars. Simultaneously, a similar program was undertaken in Japan. These efforts were focused primarily on basic infrastructure, thus repairing the foundation upon which personal and business investments could be made.

It is popularly assumed that major infrastructure change only happens in crisis. But there are also examples of such change with no imminent crisis. Between 1956 and 1970 the US built the 41,000-mile interstate highway system, while containerization was established and jet planes transformed airlines. Meanwhile we leaped from terrestrial to satellite communications, radio to color television, vacuum tubes to transistors, adding machines to mainframe computers, and from easy listening to rock 'n' roll. From idea to global usage, all in 15 years.

In 1985 computers, cell phones, digital music and the Internet were all hot new things, but they involved relatively few people in a few countries. In 2000 they were necessities involving billions of people all over the world. From, "interesting but it's too expensive," to "where do I get one right now?" all in 15 years.

In relation to our current situation the critical realization is the matter of time. It is popularly assumed the decline of ecological health and economic

Freeways to new homes, a new world invented by marketing a dream of mobility

INTERSTATE HIGHWAY
REVOLUTION: 1955 TO 1970

HOMEHOMEHOME **SUBURBAN** GROW **THEXPLODES** BY **HIGHWAY** HOMEHOMEHOMEHOMEHOMEHOMEH
MEDIAMEDIAMEDIAMEDIA **TV** FROM **A FEW** B&W **TO EVERYONE** WITH **COLOR** MEDIAMEDIAMEDIAMEDIA
TRANSPORTTRANSPORTTRAN **INTERSTATE** HIGHWAYS **FROM** ZERO **TO** 41,000 **MILES** TRANSPORTTRA
TRANSPORTTRANSPORTTRAN **CARS** ARE **FASTER** & **ROADS** OPEN **INDUSTRY** GROWS TRANSPORTTR
TRANSPORTTRANSPORTTRANSP **INTERSTATES** SPAWN **LONGHAUL** TRUCKING **INDUSTRY** TRANSPOR
TRANSPORTTRANSPORTTRANSP **CONTAINERS** TRANSFORM **GLOBAL** INDUSTRY TRANSPORTTRANSPO
TRANSPORTTRANSPORTTRANSPO **AIRLINES** GO FROM **NONE** TO **ALL** JETS TRANSPORTTRANSPORTTRAN
ENERGYENERGYENERGYENERGY **NUCLEAR** POWER **FROM** NO **PLANTS** TO **HUNDREDS** ENERGYENE
ENERGYENERGYENERGYENERGYENER **PHOTOVOLTAICS** MAKE **SATELLITES** FEASIBLE ENERGYENERGYEN
TELECOMNTELECOMNTELECOMNTELECOM **SATELLITES** START **GLOBAL** TELECOMM **INDUSTRY** TELE
SCIENCESCIENCESCIENCESCIENCESCIENCESCI **SPACECRAFT** FROM **ZERO** TO **MAN** ON **MOON** SCIENCESCIENCESCIEN
TRANSPORTTRANSPORT **TROLLEY** LINES **ABANDONED** CITIES **BUY** COMPANIES TRANSPORTTRANSPORTT
TRANSPORTTRANSPORTTRAN **RAIL** PASSENGER **THEN** FREIGHT **BUSINESS** COLLAPSES TRANSPO

1950 1951 1952 1953 1954 1955 1956 1957 1958 1959 1960 1961 1962 1963 1964 1965 1966 1967 1968 1969 1970 1971 1972 1973 1974 1975

Following World War II and the Korean War the US was poised for transformation. The country built a staggering quantity of infrastructure, all of it linked by 41,000 miles of new interstate highway. This investment transformed the landscape of modern society. These trends were revolutionary because they either initiated whole new industries, as with space exploration, or totally transformed an existing one, as in propellor to jet planes, or all but replaced an existing industry, as in interstates nearly replacing railroads.

wealth is a steady geometric descent, analogous to driving down a long grade of a mountain pass. More accurately, we're descending and the grade is getting steeper and we're going faster. Ecosystems and economic systems are comprised of a diversity of elements that have co-evolved over time and rely upon one another for survival. An ecosystem might be able to sustain the loss of one or even several species, but as the rate of species extinction increases a complex of relationships comes unglued, causing extinctions to accelerate. Similarly, a massive corporation might lose key sources of revenue, in turn triggering cuts in marketing, which results in the loss of several more important revenue sources, and so it goes. Such is exponential as opposed to geometric change.

Paradoxically, the same exponential change can occur in reverse. Just as it is possible to develop one business and indirectly trigger the formation of four more, which might then trigger the formation of another few dozen, it is also possible to restore an ecosystem, creating an exponential rise in species diversity and overall health of the ecosystem. Unlike the popular analysis of growth, where it's often erroneously assumed human population growth is the engine of economic growth, the restoration of ecosystems and

All the trends and technologies necessary to transform industrialization, mitigate the impacts of climate change, and address a raft of other problems, already exist. These initiatives need only be expanded on a larger scale. Some activities will generate results in a few years, others in 5 or 15 years. Many ecological restoration projects may evolve into more complex ecosystems over 15 to 35 years. These activities would include related initiatives, such as small town revitalization, that affect how we live and how much energy we use.

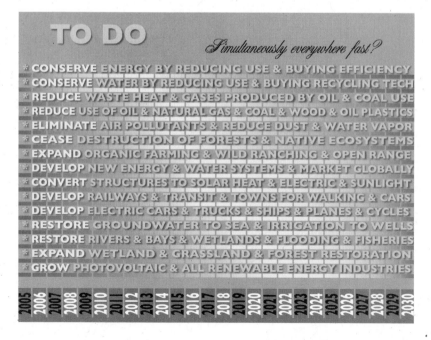

TO DO   *Simultaneously everywhere fast?*

* CONSERVE ENERGY BY REDUCING USE & BUYING EFFICIENCY
* CONSERVE WATER BY REDUCING USE & BUYING RECYCLING TECH
* REDUCE WASTE HEAT & GASES PRODUCED BY OIL & COAL USE
* REDUCE USE OF OIL & NATURAL GAS & COAL & WOOD & OIL PLASTICS
* ELIMINATE AIR POLLUTANTS & REDUCE DUST & WATER VAPOR
* CEASE DESTRUCTION OF FORESTS & NATIVE ECOSYSTEMS
* EXPAND ORGANIC FARMING & WILD RANCHING & OPEN RANGE
* DEVELOP NEW ENERGY & WATER SYSTEMS & MARKET GLOBALLY
* CONVERT STRUCTURES TO SOLAR HEAT & ELECTRIC & SUNLIGHT
* DEVELOP RAILWAYS & TRANSIT & TOWNS FOR WALKING & CARS
* DEVELOP ELECTRIC CARS & TRUCKS & SHIPS & PLANES & CYCLES
* RESTORE GROUNDWATER TO SEA & IRRIGATION TO WELLS
* RESTORE RIVERS & BAYS & WETLANDS & FLOODING & FISHERIES
* EXPAND WETLAND & GRASSLAND & FOREST RESTORATION
* GROW PHOTOVOLTAIC & ALL RENEWABLE ENERGY INDUSTRIES

2005 2006 2007 2008 2009 2010 2011 2012 2013 2014 2015 2016 2017 2018 2019 2020 2021 2022 2023 2024 2025 2026 2027 2028 2029 2030

the economic value such restoration may spawn is not inextricably linked to our population. It represents a shift in the sources, quantities, types and values of species, both animals and plants, not necessarily in the quantity consumed by people. People eat lots of burgers in Alaska, only it's a mooseburger not a cowburger.

Given human history, specifically the astonishing things we've built in a very short time, plus the very idea of positive exponential growth, it would seem quite possible to transform infrastructure within 15 to 20 years, and transform the world economy and culture at the same time. Such a boggling possibility may seem absurdly optimistic, yet history reveals we've done it before, and now we possess a stunning array of new technologies that make it both more feasible to consider and more practical to do. It could very well be a future as bright as it is now dark. It is just possible.

In forming a vision our brains are a buzz of electrical activity as we concoct a picture of the future. It may be the vision of an event in a few weeks, a building in a few years or a whole city in a decade. The vision is a buzz of tiny electrical charges between cells, all within a cradle of water. A vision of the future is defined by prescience of unknown origin, driven by our faith in the future we want to create. It's a thought about earth, air, fire and water. It's a picture of a new world born in water like all of us — in light of water.

2

A Wealth of Energy and Water for All

Don't Unplug my Life

WE ARE ENERGIZED WATER. Without sufficient water and energy, modern industrial civilization, along with any hope of improvement in the quality of people's lives, would dry up like a puddle in the sun.

Among those who study such issues, and evidenced by numerous articles in prestigious publications, it is widely believed global water and energy sources are taxed to the limit, and therefore our future as civilized nations is dubious. There are already severe water shortages in many areas of the world, and several oil-producing regions are within a decade of being exhausted — notably Mexican oil fields. But the public debate over these issues is remarkably shallow, all but devoid of any substantive understanding of how existing systems work and the new technologies already available.

Long before global warming registered on the political thermostat, as far back as the 1960s, many scientists and industry leaders in the world of water and energy were beginning to develop new technologies. The technologies now emerging as real products represent a potential future where there is ample energy and pure water available to everyone on earth. In this future there is no need to maintain a military presence to guard oil fields because everyone has the energy they need and no one controls the supply. In this future pollution is vanishing, climate change is diminishing, ecological wealth is growing, economic wealth is expanding and human populations are stable or declining because the quality of life has improved for the majority of the Earth's people.

Technologies already on the market set the stage for one system that supplies electricity and pure water, and one system for powering buildings and vehicles. This system, plus somewhat more sophisticated appliances in homes and commercial buildings, will replace the long-distance grids, aqueducts and piping systems, as well as dams, coal mines and oil fields. These new water-energy technologies are equivalent to the miniaturization of all the technologies that comprise the massive utility systems we now rely upon. All that technology can now be reduced to a unit about the size of a home furnace helped by a few modest changes to roofs and walls and a few new appliances, changes so thoroughly integrated with buildings the untrained eye wouldn't notice anything unusual. Cities would no longer need power plants or distant reservoirs; they'd be the power plant and reservoir. We would rely on sunlight and rain.

Amazingly this future is in development now, and it's even represented in advertisements on television and in magazines. There are advertisements in major newspapers for home-scaled solar roofing technology, and there have been full-page ads in magazines like *Atlantic Monthly* and *Harper's* for General Motors' hydrogen-electric automotive program and BP's solar-photovoltaic division. Yet the potential of these and many other technologies has until recently been almost totally ignored in the mainstream debates about energy and water. How could such a remarkable possibility be unseen, like an animal hiding in plain sight?

Contemporary society, rife with visions of Armageddon, is blind to this bright future. Five decades of cold war fear killed faith in the future, leaving a society with a pathological tendency to assume the worst. Popular visions portend a grim world where we humans are either fighting endless resource wars over water and oil as the long hot days toast the temperate zones, or are being destroyed in a horrific final war with the chosen people blissfully experiencing rapture on their way out.

The popularity of Armageddon fantasies, whether presented as a science fiction warning of impending doom or as a religious fantasy of ultimate transcendence, represents a dominant cultural feeling of powerlessness over one's future, and often a profoundly cynical excuse for inaction. Proponents of such grim fantasies thrive in a nation where expectations of a better future are quietly vanishing along with real buying power. A horrific vision of the future serves politicians who seek power by fear, and individuals who seek to justify their inaction by claiming nothing matters because the world is coming to an end anyway.

Invariably such visions of doom are built upon the idea of scarcity, a scarcity of water, energy and food leading to a scarcity of civil behavior. It's widely assumed that there isn't enough energy and water for all people, and therefore the quality of life that defines modern civilization is unattainable for much of the world's population. This is the premise of scarcity and the basis of a profound fear underlying all manner of conflict around water and energy.

Beliefs about scarcity are largely based on a quality of life defined in US terms, and a shallow analysis of existing and new technology. In fact, numerous societies in the world today have a quality of life that is as good or better than the US standard, yet they use far less of everything. If technology already in use were widely applied, we could cut US energy and water consumption by more than half and no one would notice any difference, except that their utility bills would be lower.

The popular response to scarcity is typified in two extreme visions: The eco-vision where everyone recycles, gets by on a meager allotment of solar-energy and drives tiny cars, or better yet rides a bike, and the techno-vision where nuclear fusion provides ample power for the right people who can afford it because they won the geo-political game of controlling the enormous quantities of vital resources necessary to build nuclear power plants.

These two visions are cartoons revealing a grotesque misunderstanding of the technologies we now rely on. The most extreme eco-visionaries tend to assume energy and water are in short supply and *always* will be, that cars are intrinsically unsustainable and that transporting food long distances will become impossibly expensive as energy prices rise. The problem is this vision is built on dubious assumptions. The cost to move most commodities, especially by rail and ship, can be as little as one percent of the retail price of the product in the store. Wind turbines, solar-thermal and photovoltaic technology, plus a smattering of variations, could generate ample electricity worldwide within a few decades and all evidence suggests this will happen in direct proportion to a general rise in energy prices. Emerging water technologies and management strategies in use by major cities suggest it's possible to radically reduce reliance on imported water. Ample research on new materials, and breakthroughs like Ford's electric soy-based plastic concept car, suggest the potential of cars that could be manufactured indefinitely using materials grown in fields and the energy of light — sunlight.

Solar energy is often discussed as if it were somehow insufficient to meet our needs, especially for heavy industry. Such assumptions border on the absurd when one considers the staggering quantity of energy we're receiving at any given moment.

ENERGY FROM SUN STRIKING EARTH IN ONE HOUR

HOURLY SOLAR INPUT = 19,000,000,000,000 WATTS /ANNUAL GLOBAL CONSUMPTION=13,000,000,000,000WATTS

ENERGY USED IN ALL HUMAN ACTIVITIES IN ONE YEAR

The most extreme nuclear visionaries tend to assume the opposite of the eco-visionary position, yet they too build their vision on very dubious assumptions. They assume a power plant is the necessary mode of generating electricity; that nuclear is the best option because it generates no greenhouse gases; that it makes sense to use lots of non-renewable energy, primarily oil, to build thousands of such plants; that irradiation of a vast area of the former USSR constitutes a good safety record; that one day someone will figure out how to store radioactive waste; that one day someone won't figure out how to make a bomb; and that the private sector *will* invest, someday.

It would seem, as modern society climbed the ladder of progress and reached the high-tech rungs represented by computers, cell phones and the Web, we lost sight of the bottom rungs. Few people recognize that much of the electricity that energizes their computers and chills their veggies comes from power plants burning coal received by rail from mines, the same infrastructure that initiated industrialization 150 years ago. We've forgotten everything at the other end of that wall socket or faucet.

Fear permeates contemporary debates over energy and water. Fear of terrorism is in part couched not as fear of some political philosophy that is so compelling it causes true believers to fly planes into buildings, but as fear of chaos because "they" can attack an infrastructure we no longer understand. Fear over US vulnerability to a terrorist attack on our energy supply has not lead to defensive action to reduce consumption and achieve self-reliance, but to offensive military action to secure shrinking oil supplies.

We're Drenched in Energy and Water

Any debate on water or energy can become Byzantine in its complexity simply because the subjects are both technically arcane and extremely potent in their implications. Worse, energy and water are the subject of endless machinations orchestrated by all sorts of special interests, from companies seeking control of city water supplies to water districts bent on assisting political friends with a lavish pipeline to a previously worthless desert. Invariably such intense lobbying distorts the perception of energy and water as a litany of simplistic ideas and opinions designed to suit special interests with no awareness of what the general interest is. All too often the debate is resolved not by any agreement over general interest, such as the innocently brilliant objective of providing pure water and clean energy for all, but by whoever produces the best public-relations campaign.

Special interests often succeed in making changes they claim are in the best interests of all, only to improve their own standing at the expense of all. There are, for example, those who see price as the only measure of value. In the 1990s these free-market advocates succeeded in creating a North American market in electricity by causing what were once independent regional

In the US the majority of water is delivered by government agencies and public water companies via reservoirs. Nearly half of this water is used in boilers to make steam, which drives turbines to generate electricity. In some farming regions wells are a significant source; about a third of agricultural irrigation water in California's Central Valley is groundwater. Industrial water is generally used for cleaning and manufacturing processes, and as a source for various food products. (Source: United States Coast & Geodetic Survey)

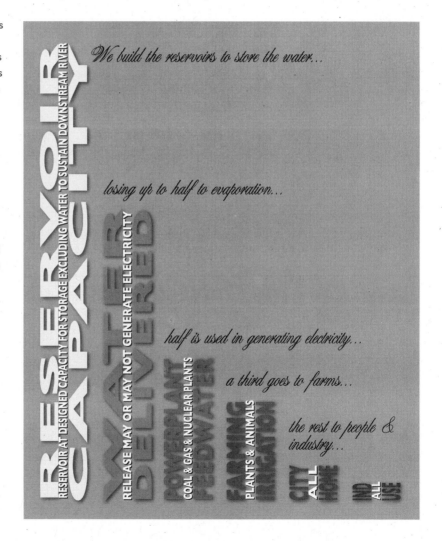

electrical grids to function as one marketplace — a continental grid. Their argument: lower prices for all. Whether the energy was dirty or clean didn't matter.

In the same period the nation became more dependent on imported oil because too many voters saw only the price at the pump, not the price of defense expenditures in oil-producing regions. In effect, we've sustained the illusions of low prices and national security while in fact reducing the nation's competitiveness by continued reliance on oil that requires a costly

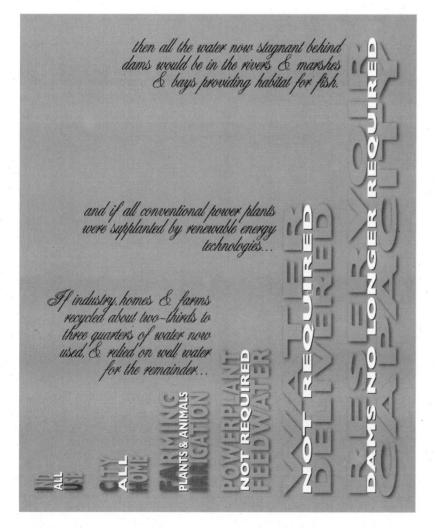

then all the water now stagnant behind dams would be in the rivers & marshes & bays providing habitat for fish.

and if all conventional power plants were supplanted by renewable energy technologies...

If industry, homes & farms recycled about two-thirds to three quarters of water now used, & relied on well water for the remainder...

INDUSTRY ALL USE
CITY ALL HOME
FARMING PLANTS & ANIMALS IRRIGATION
POWERPLANT NOT REQUIRED FEEDWATER
WATER NOT REQUIRED
DELIVERED
RESERVOIR
DAMS NO LONGER REQUIRED
CAPACITY

If the majority of water used in the US were supplied by recycled water and groundwater via wells, and assuming a conversion to renewable energy, nearly all the water now being stored in reservoirs, with up to half lost to evaporation, could be returned to the rivers, thus allowing restoration of fisheries and downstream groundwater.

military guard. We've also increased our vulnerability to natural, accidental or intentional disruptions of electricity by reliance on a delicate grid. Failure is the price of ignorance.

A society of specialists tends to see single solutions to single problems, and this tendency guarantees scarcity of energy and water. This is especially visible in relation to water.

Most of the US now relies on pure water piped great distances from remote reservoirs. We use a mere one percent for drinking, while the rest is used for washing and landscaping. In most cities dirty water, along with water collected by streets and roofs via storm drains, is piped to tertiary (third stage) sewage treatment plants where it's cleaned to 98 percent-plus purity, only to be dumped into the watershed to flow to the sea. Since used water is disdained like a pre-owned car coastal cities are now building desalinization plants to get more water!

Let's get this straight: we move pure water a great distance because it's pure, we barely drink any and dirty the rest, then pipe it some distance to a factory that virtually purifies it, only to drain it into the ocean and then build a plant to take the salt out? Then millions of us buy bottled water, at several times the price of gasoline, because we don't trust the tap water.

Wealth by All Measures

New technologies already on the market could result in everyone on the planet having ample water and energy. Nothing need be invented.

We are saturated in energy and water. Imagine a field extending as far as the eye can see and covered in 100-watt light bulbs one foot apart. The power consumed by those bulbs equals the solar-energy that falls upon the earth. Rain falls upon virtually all our cities at the rate of 10 to over 100 inches a year, depending on where the city is. This energy and water is freely distributed to virtually all people, whether they want it or not.

What if one could buy a new form of appliance, an energy-water system in a box, that would replace water heaters, furnaces and air conditioners and rely on sunlight and rainwater already delivered? What if this system could be manufactured in different sizes and sold through home hardware stores at mass-market prices ranging from $10,000 to $25,000 or roughly equal to the average kitchen remodel? What if millions of such energy-water units replaced existing water, electricity, gas and gasoline systems, offering pure drinking water and clean energy, while removing all pollutants from air and water? Such an appliance does not exist, but it is possible to build one now.

Such an appliance would change the world. It would capture heavy metals, chemicals and drugs, steadily purifying the waters of Earth. It would end air pollution caused by fossil fuels, decreasing healthcare costs. It would end reliance on fossil and nuclear energy, thus allowing a reduction in military expenditures. It could end reliance on conventional power plants and large dams, allowing restoration of watersheds and fisheries, and it would redefine the idea of energy independence, from national to individual independence.

Using building technologies and solar technology available today it's possible to build a structure that provides sufficient electricity, water and heat to support one person using a ten by ten foot roof; assuming about 20 inches annual rainfall, with 75% of their water recycled, and excluding garden water, which varies widely by region. They would need a water tank up to about 5 feet square. A typical household would be about 3.5 times larger.

The implications in the developing world are even more profound. What if most of the world's population, within the next 20 to 30 years, could have ample energy and pure water? How many children would survive because their water was pure and refrigeration led to the ability to store food and medicine? How many millions of people could develop a business because they had the power to reach the world? Could this one change cause a global economic renaissance measured in the trillions of dollars? Could one family of technologies, if expeditiously developed, stop and reverse climate change trends?

Roofs are perhaps the most under-utilized asset in the world. A small L-shaped suburban home roof might be about 1,700 square feet. It could provide energy, water and heat for the average 3.5 residents of the average dwelling unit, of which there are about 101 million in the US. The roof can capture upwards of 21,000 gallons a year at 20 inches of rainfall and this, plus water absorbed by the ground and collected by streets, could provide ample water for homes if 70 to 85 percent of usage is recycled.

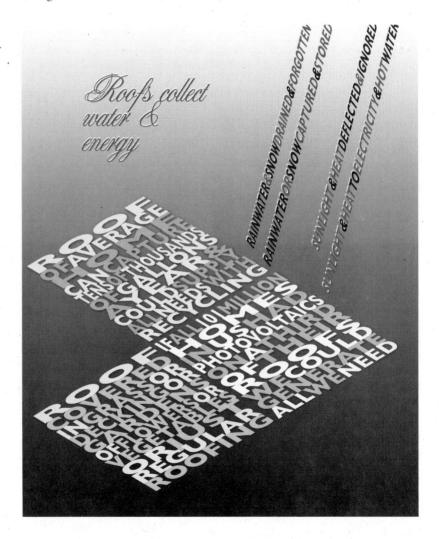

Fantasy or reality? No doubt many people would dismiss the notion as delightfully utopian. In fact the energy-water system is represented by technology long used in the US space program. In space the shuttle, the international space station and all manned spacecraft are effectively powered by the sun, with photovoltaics providing the electricity, hydrogen used to store the electricity and all drinking water endlessly recycled. A few buildings now feature similar demonstration systems, but as yet no one has assembled a system packaged for mass production. Nevertheless, it's very likely such an energy-water system will be developed because the technologies that make it possible are already produced by some of the world's largest corporations in the energy, automotive and semiconductor industries. In their reality the business of new energy and water technologies represents a market nearly equal to the population of Earth. In our reality it will likely happen because we'll just pay our utility bill, but get something different.

It will also happen because it will be profitable by every measure, offering benefits in cultural, environmental and monetary terms. In relation to any new technology it is widely believed price is the only measure of feasibility and financial profit the only measure of motivation. Were this true there would be no modern industry.

Regardless of the most cynical view of humanity's prospects, as expressed in your Armageddon of choice, people do make decisions now about a future they cannot know by developing alternatives. As a society we've been developing alternative energy and water technologies for a few decades. The work wasn't based on short-term profit because for most of the pioneers profits were ephemeral. It was based on the idea of profit by all measures: quality of life, a legacy to one's children and maybe a good living doing good work.

Remarkably, despite a wealth of innovations in renewable energy and water purification technologies there is little popular awareness of the implications. Instead popular debates remain focused on getting energy and water from elsewhere. It is as if we were standing in the rain dying of thirst, or standing in the sun freezing to death.

The only shortage of energy or water is in the mind. With a modicum of technology available on the world market everyone could have what they need.

3

Less is More of Something Else

Consuming to Disaster

SOCIETY HAS A SPLIT PERSONALITY when it comes to resources. An employee of an organization is told to do more with less, but when the same employee goes home they are a consumer being urged to do more with more.

For over a century virtually all businesses sought to use less of all resources while encouraging customers to use more resources — their resources. Companies sought to become more efficient in order to achieve an attractive return on investment and survive. Every penny wasted is added to costs and subtracted from profits, and if profits are insufficient to sustain investment the business is over. This imperative has resulted in staggering efficiencies. In 1860 a locomotive generated 16 horsepower to move one ton, now a locomotive generates one for one.

Unfortunately there remains a widespread belief, created by decades of unrelenting advertising, that consuming is success and conserving is failure. This belief drives endless political initiatives to get more water and energy no matter the cost. Yet there's ample evidence that innovations already in existence allow dramatic reductions in energy and water consumption with no loss to anyone's quality of life. In most cases a conservation investment is

easier to do, and pays back its cost in savings in less time compared to investments in new generating capacity. Conservation is a benefit not a penalty.

By the Light of Stars

Lights are a major consumer of electricity. We turn on the light without
even a second thought about what it means to turn on light. The light bulb
has gone from the miraculous to the mundane in 125 years.

In the last few decades the light bulb has been undergoing revolutionary
change. Average contemporary bulbs are already much more efficient than
bulbs of 25 years ago. New bulbs are about to cut that in half, specifically
new light emitting diode (LED) bulbs now on the market. These new lights
can be so small a reading light would resemble a diamond-like star in middle
of a room, so precisely color-balanced they resemble sunlight, yet so efficient
they use less than half the energy of contemporary bulbs.

LED bulbs, unlike conventional bulbs with filaments or fluorescent gas,
can be turned off and on with no damage to the bulb. This has profound implications. In 2006 the Herman Miller furniture company introduced a
lamp called the "Leaf." Its tiny LED bulbs take a cue from the movie business. Movies are based on the fundamental fact that the human eye per-
ceives still images as continuous because the images are flashing at 30 or
more frames per second. The lamp's tiny bulbs turn on and off so fast the eye
sees continuous light. This simple idea cuts the lamp's already low power
consumption by more than 40 percent.

New technologies, particularly light emitting diodes, have exceptional
implications for reducing energy consumption, largely because lighting represents a large portion of total energy consumption. But it must be said that
too much lighting, as now visible nightly in any urban area in the world, is
perhaps the single most wasteful use of energy and money on the planet. At
any given moment innumerable signs and displays are illuminated in midday while empty rooms or whole buildings are flooded with light when no
one is present. Leaving lights on to deter criminals may make one feel better,
but it isn't going to deter any serious burglar. What reason could there be for
leaving so many lights on?

Streetlights have become a salve for the fears of many and a staggering
waste of energy. In the last few decades the US has become a nation where

more light is equated with less crime, hence a manic attempt to mimic sunlight at midnight. From 35,000 feet, looking down on the civilized world at night reveals a landscape of empty parking lots and highways as bright as day.

We need light in the night but it needn't be so bright. We could cut the wattage by more than two-thirds. Less light would make no difference in terms of crime as any serious criminal isn't going to let streetlights spoil their evening. A large percentage of streetlights could simply be turned off.

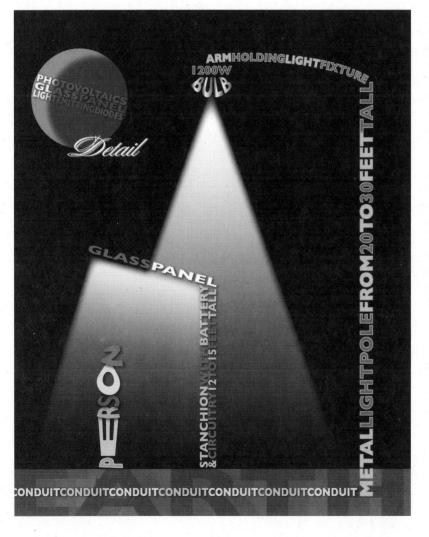

Existing streetlights waste energy and blind us to the night sky. The potential of new lighting strategies is characterized by a new Sanyo streetlight (available in Japan). It has a battery in the support and a glass panel incorporating photovoltaic and light emitting diode films (detail). It would receive daylight, generating and storing electricity, later providing a glowing light at night, instead of a glaring spotlight. If motion sensors were used lights would only be on when someone is present. Where crime is a concern savings from streetlight energy costs could be used to hire more police. No conduit or power plant required.

Others can be fitted with motion sensors to turn on lights as needed. The very presence of motion sensors would deter most criminals, or announce their presence.

Perhaps it would make more commercial sense to develop streetlights that offered a softer glow, and made people want to be outside. What if there was a short streetlight, with no bulb, just a panel of glass emitting a soft warm light over the sidewalk? During the day the same panel of glass, the dimensions of a small window but unframed, would seem almost invisible but for its slight tint. The glass, coated with microscopic LED bulbs on the bottom and a photovoltaic film on the top, and supported by a vertical stanchion containing a battery, would receive solar-energy during the day and glow at night. Sharp Corporation makes just such a product for the Japanese market.

Darkness needn't be a source of fear, it can be a comfort and a signal to the body to relax. Maybe one day we can get back the moon and the stars. In the meantime, turn off outdoor lights on moonlit nights.

Sunlight is the standard by which all light bulbs are measured because sunlight is full spectrum color. There is evidence to suggest we are at least more comfortable if not more productive in sunlight. Sunlight, specifically ultraviolet light, causes our bodies to produce more vitamin D, vital to our immune systems. Indeed, for our health and emotional well-being, plus our economy, it would seem vital we have more sunlight *and* more darkness.

Skylights are sky windows. In early 20th Century large structures often incorporated skylights as a means of reducing the cost of lighting. Factories often had a sawtooth roof, with angled roofing facing south and vertical windows facing north, thus flooding the factory floor with indirect daylight. But as brighter bulbs became more feasible, often leaky skylights were replaced by fluorescent lights. Cheap lighting made it possible to illuminate any space, in turn making it possible to build acres of tilt-slab industrial buildings with no windows. Could it possibly be healthy to live or work in windowless rooms?

Sky lighting has made a comeback. Improved designs have made skylights less costly and, best of all, less leaky. Even glass roofs are more feasible. Increasingly, new homes and commercial structures include skylights that flood a room with sunlight, often eliminating the need for electric light during the day. Wouldn't many people feel better reading a lower utility bill in a room filled with sunlight?

Light pipes exist. Every day millions of people talk on the phone or send messages via the Net, transmitting their words and pictures through light in tiny fibers of glass. The same fiber-optic cables could carry sunlight. A lens atop a pipe can focus sunlight on a bundle of fibers. The light beams, contained within the hair-thin fibers like water in a pipe, would flow down the fibers to small fixtures set in the ceiling where the soft glow of sunlight emanated from the polished ends of thousands of fibers. A windowless room could be filled with sunlight.

New LED bulbs will be merged with fiber-optics to create lighting fixtures. A subway station in Manhattan could be filled with sunlight during the day. As the sun went down sensors would turn up the artificial light and those waiting for a train would notice the transition from the pinkish glow to a warm white incandescent light. On a clear winter day Fifth Avenue stores would be in shadow all afternoon, but store windows would be lit by the afternoon sun still striking the roof of the building.

Keeping Heat In or Out

Insulation is boringly silent, yet it's as vital as a blanket in the Arctic. Insulation is increasingly a design strategy as well as a building material. Combine new wall materials and new glass and a building can be warm inside when it's icy outside, or cool inside when it's baking outside. Multipane windows can be so insulative one can have a picture window view of a polar bear in a blizzard from a cozy warm living room on the edge of Hudson Bay.

Recent revolutionary changes in insulation represent a major leap in the sophistication of measuring how energy moves. Architects and engineers can now design spaces that require less energy because they retain or remove heat more efficiently. This understanding has led to new materials, such as plastic films as thin as paint with the insulative value of a foot-thick adobe wall. Using available technologies it is possible to reduce a home's energy consumption for heat to near zero — an occasional wood fire on a few cold days — and rely on solar gain, either direct via windows and/or from hot water systems, plus appliance heat and the heat of people. This can and is being done, even in Arctic regions north to the midnight sun.

Contemporary refrigerators are more efficient than 1990 models, yet new insulative materials and the use of microprocessors to control refrigeration systems promise even further improvements. Refrigeration works in part by the use of materials in the surrounding box that do not transmit heat. Such seemingly static and mundane materials have become exceedingly sophisticated as scientists have uncovered the elegantly simple realities of how molecules transfer heat. Their work has led to insulative materials that may be nothing more than a cloud-like gel, a film so thin it's barely visi-

Existing building codes in North America ensure that more recently built homes and commercial buildings are energy efficient. Most older structures are energy sieves, losing heat-bearing moisture from cracks and by transference via single pane windows. Conventional incandescent bulbs, older appliances and excessive use of appliances results in a significant waste of electricity. By using high efficiency appliances, insulation, vapor barriers, multipane windows and natural lighting most structures can be cooled with little or no energy consumed, and warmed by appliances, people and the sun.

ble, or a "smart" film with molecules that automatically realign themselves to admit or repel heat. Given that refrigerators are a major and constant consumer of electricity, and one vital to our health, efficient refrigeration is no trivial matter.

Air conditioning is a form of refrigeration. It is also symptomatic of buildings that don't fit Earth. Until recently most people assumed a building was a building regardless of whether it was in Ottawa or Orlando. Commercial building design was driven by the economics of leasing. Heating and air conditioning costs were an operating expense and weather was irrelevant. Typifying this disconnection most high rises consume more energy to reduce heat generated by lighting, appliances and the sun than they do to electrify the appliances. Air conditioning is equivalent to paying to use fossil and nuclear energy to remove renewable energy we receive anyway, for free.

Several US cities are involved in a federally sponsored "Heat Island Initiative" aimed at reducing the heat produced by cities from buildings and pavement. In Houston the average summer temperature downtown and in suburban office complexes was an astonishing six to eight degrees warmer than in surrounding rural areas. The local program is called "Cool Houston" and it focuses on a variety of strategies including trees, permeable pavement and parking lots "paved" with grass. Building owners often see major savings in heating and cooling costs by using trees and green roofing.

Energy is a study in paradox. It was once thought solar homes had to look like cracker boxes. Architects were so focused on making the building into a solar collector they didn't notice it was also collecting the heat generated by appliances and inhabitants. But after a few decades of experience, and with new insulative technologies that retain heat when it's cold outside or resist heat when it's hot, it is now possible to develop structures that remain at a comfortable temperature. Such structures don't need furnaces or air conditioners because they *are* the furnace and air conditioner.

A growing number of European buildings have green roofs. Several years ago architects rediscovered the ancient idea of a sod roof. It turned out to be a very efficient way to reduce noise and heating costs. Today there are hundreds of buildings with roofs sprouting grass on a thin mat of soil laid atop a watertight roof. The soil and blades of grass effectively store moisture that helps insulate the structure. Plus it looks cool, a home with a buzz-cut roof, or perhaps a Mohawk.

Energy, Water and Conservation by Design

Energy and water are inextricably linked. Hydroelectric dams not only contain water for urban and agricultural use, they provide electricity. Water systems are among the biggest users of electricity, both for pumping water over mountains and powering urban treatment plants that purify water. Energy systems, specifically electric power stations, use nearly half the total freshwater consumed in the US as feed water for boilers generating steam to drive the turbines that generate electricity.

The energy conservation trend of the past few decades has been paralleled by trends in water conservation. Water agencies, scientists and appliance companies have made breakthroughs in all aspects of water usage — a water tech revolution. Since the seventies public water agencies have instituted significant conservation strategies as they found themselves caught between a lack of available reservoir sites and a public increasingly opposed to further loss of what little remains of wild rivers. The strategies vary by place and weather, but the trend is focused on the three "Rs": reduce, recycle and restore.

Reducing use means developing techniques and technologies that require less water. Technologies like low-flow toilets, timed faucets in public restrooms and pool covers can all reduce personal water consumption dramatically. Contemporary landscaping and many crop irrigation strategies involve drought tolerant plants, drip irrigation and even simple changes like using sprinklers at night. In many arid regions, landscapers have developed "xeriscape" strategies for homes and public spaces to reduce or eliminate the need to water plants and build costly sprinkler systems. Xeriscaping favors plants adapted to survive dry periods. They often use native plants.

Recycling means purifying and reusing water instead of just draining it to the sea. Water can be partially recycled within a building, where all toilet water passes through a sewage removal system and is recycled as "greywater." It's actually crystal clear, not gray. It can be recycled by pumping water from sewage treatment plants underground, where tiny spaces in stone becoming a water filter and a water bank, with withdrawal by the same pump.

Commercial-scale recycling systems have been developed in several large buildings.

The Los Angeles Department of Water and Power (LADWP), an agency whose water grabbing political history was characterized in the movie "Chinatown," now supports a host of new water-related initiatives. Nearly a century ago, under circumstances of dubious legality but brutal intentionality, Los Angeles secured the water rights to the remote Owens River watershed and built a long aqueduct. They began to drain the region of its water, interrupted a few times by enraged locals with dynamite. First

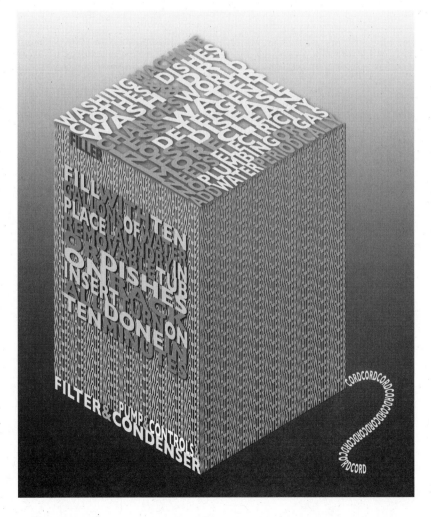

Existing water systems move countless gallons of pure water from distant reservoirs, all so the water can be mildly dirtied and drained away. This makes no sense. A new washing machine, potentially for dishes and clothes, could not only do the washing in less time, but store ten gallons of water internally and clean it — no plumbing required. Once a month or so, depending on usage, one would add a cup of water and remove the filter. Lint and food waste would be recycled as fertilizer or bought by a recycling company.

the Owens River and Owens Lake dried up, and then the city began drain-ing deep aquifers. Finally, in a series of judgments in the 1980s and 1990s, federal courts sided with the residents and forced LA to stop draining the aquifer and restore the river and the lake. Some restoration has been done, but as of 2005 LA and Owens Valley counties were again in court.

In response to the loss of some Owens River water the LADWP, along with other Southern California water agencies, have developed aggressive conservation strategies, including incentive programs to encourage home-owners to conserve, as well as industry programs to encourage recycling. Meanwhile the agency is pumping water from sewage treatment plants un-derground, developing rainwater catch basins and becoming involved in ecological restoration.

New technology, especially smaller electric motors, electronic controls and filtration systems, suggests even further reductions in water and energy consumption are on the horizon. These new technologies invariably involve seemingly tiny improvements in existing technologies, such as faucets with ceramic gaskets that do not leak, or applications of elegantly engineered membranes to filter water.

A new view of water may well yield new products. Dish and clothes washers are boxes with an inner box or cylinder containing the dishes or clothing. Between the inner box and outer box there's an electric motor, some hoses, mechanical parts and a fair amount of air. By redesigning the machine, this space could become a water tank. Moreover, the clothes wash-ing machine could also be a drier, and perhaps a clothes and dish washer-drier could be developed — one unit instead of three. Water could be end-lessly recirculated and filtered. Once a month a filter cartridge would need to be removed, with its compressed waste and lint, and perhaps a cup of water added to replace that lost to moisture on dishes and in clothes.

It is possible to develop a toilet, even a whole bathroom unit with sink and shower, that does not require water or sewer connections — only elec-tricity and several gallons of water. It would look and flush like a conven-tional toilet, but it would be a clean box shape, much like a bench. The water would be recycled and the waste would be dry odorless pellets encased in a biodegradable film — fertilizer. Models with a heated seat and bidet func-tion could also be available.

New water appliances could cut home water consumption by more than two-thirds. Piping systems would be much simpler, with no need of water pipes to all appliances and often no sewer connections in rural and suburban areas. Centralized water heating, with its hot water piping requirements, could be replaced with precise on-demand heating by heaters in faucets. No more waiting and wasting for hot water to arrive.

New technologies and design strategies that use less energy and water are often perceived as just that: pragmatic options to become more efficient and save money. The very notion of conservation suggests a certain miserly attitude. But the reality of new technologies is far more compelling than just lower utility bills. New lighting, combined with skylights, means higher quality color-balanced lighting, more sunlight and more vitamin D. New insulation translates to a quieter interior, and a reduction in the general background hum of modern cities. In short, quality of life goes up and cost goes down.

Landscaping with native plants characterizes the larger implications of conservation. Deciduous trees adapted to a particular region can shade and cool buildings in summer, yet lose their leaves and admit maximum sun in winter, thus reducing our energy needs with no work being done by anyone or any machine. Native plants can not only survive without our watering, they also attract native species of insects and birds, often reviving a variety of other species that might have otherwise vanished, thus reinforcing the web of life that surrounds us instead of killing it. Conservation strategies point the way to restoration of the planet, and the recognition that we, all humanity, are not apart from, but a part of nature. We need birds, bees and trees.

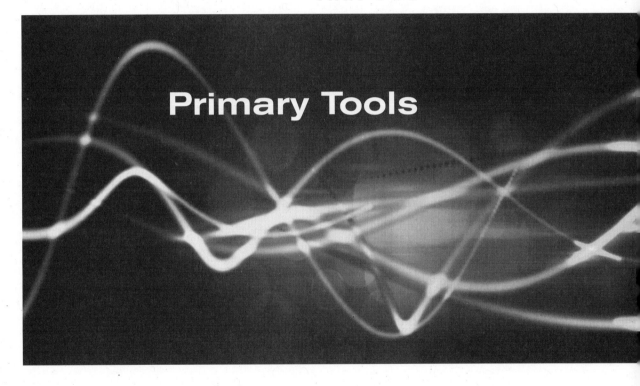

PART TWO

Primary Tools

4

Power from a Crystal as Thin as Paint

We Are the Sun's Children

WE ARE SOLAR-POWERED. We are energized by the sun, augmented by a comparatively tiny quantity of fossil fuels, plus an even tinier quantity of uranium. If the sun went out all life on Earth would end within hours. Fortunately the sun has perhaps a few billion years of hydrogen fuel, and that's more than enough to support all of us here on Earth for a long, long time in a galaxy that will be far from today. Solar power rules.

Our relationship with the sun and moon is deeply ingrained. Our behavior is timed by their movements. On full moonlit nights emergency rooms are often overwhelmed with accident victims because many of us go a little crazy under the moon — especially men. Women's menstrual cycles, where there is less artificial light, can be in synch with the light of the moon.

The quantity of light that falls on earth in one minute is equal to the total energy consumed by civilization in one year. We receive approximately 350,000,000 terawatt hours of energy (that's 350 billion trillion watt hours) every year delivered free courtesy of the sun. We only need about 100,000 terawatt hours to equal current and projected demand for all forms of energy; oil, gas, heat and electricity. That translates to roughly one-third of one percent of solar radiation income to provide sufficient energy for all citizens of Earth. If we valued earth's solar income at contemporary rates, per kilowatt hour, it would be worth in the neighborhood of $17.5 quadrillion dollars.

Einstein defined sunlight as packets of photons streaming from the sun, but there is more to the sun's light than even the most sophisticated explanation of what it is. Sunlight also drives the weather, a singular manifestation of clouds, water, air and land all acting as one continuously sustained force. Weather is wrapped around human consciousness in the vagaries and inevitabilities of crops and seasons, storms and long summer days, realities that have brutally governed our survival, yet guided the passage of our ships and our visions. Sunlight powers our story.

The energy and life of the planet is often abundant at the edges, where land meets the sea and freshwater meets saltwater. Predictably these power centers drew people to settle nearby and take advantage of the animal life, gentle breezes and currents of river and sea. Most of the world's cities are aside bays or rivers; nearly all are located where sunlit days are frequent.

Sun & sea, wind and river, endlessly exchanging energy by the mouth of the river

Electricity is not a thing made but a state created. Generating electricity means to excite electrons in their orbits around atoms so much that they leave their orbits and move on. They are so excited they will flow through a conductive wire. If there are too many excited electrons and the wire is too small and not very conductive there is resistance and the electrons, in their attempts to move on, will throw off heat. We see the result in light bulb filaments and electric stoves.

Electricity is the ideal means of transferring energy because it can be controlled precisely at all scales. Electric motors range from tinier than the head of a pin to the size of a car. Locomotives, and increasingly ships, are diesel-electric, using electric motors to drive wheels and propellers. Electricity is the means of powering lighting, refrigeration, industrial machinery and appliances. Electricity could power all vehicles, thus eliminating gas and diesel engines.

Electricity, light, magnetic and gravitational fields all permeate the lives of all living things, organizing our energies by the pulse of the universe. Electricity is also the power of mind. We think by the excitation of electrons.

Light and Heat Delivered Free

Renewable energy is a family of technologies, including photovoltaics, wind turbines and solar-thermal devices, that generate electricity and/or heat from sun, wind and water. Hydroelectric dams, tidal power and ocean thermal gradient plants are also renewable technologies. However, large-scale hydroelectric facilities are problematic due to impacts on river ecosystems. Tidal and ocean thermal facilities are limited by the number of possible sites. Wind turbines are widely applicable and there are millions of untapped sites. The most practical form of renewable energy is the photovoltaic (PV) cell. It works wherever the sun shines — nearly everywhere.

Government agencies in the US and other countries track the quantity of electricity generated and the source of that electricity, whether it was generated by coal, natural gas, nuclear, oil or large-scale hydroelectric, or renewable sources such as wind, photovoltaic or solar thermal facilities. But these estimates count only facilities connected to the electrical grid and electricity generated by established utility companies. In the US they do not count an estimated 250,000 homes receiving most if not all their heat from the sun, either because the homes still retain a connection to the gas or electrical system, or because they are off the grid entirely, as are the 20,000 or more homes and small commercial installations that receive all their electricity from the sun. This statistical oversight yields numbers that underestimate the contribution of renewable energy, while discounting the reality and the value of independently powered buildings.

PVs are the fastest growing source of energy in the world, with current annual growth exceeding 40 percent. In order of market share in 2006, the industry is comprised of Sharp, Q-Cells, Kyocera, Sanyo, Mitsubishi,

Cities are widely viewed as places that draw energy and water from elsewhere. Industrialization, fueling the rise of the modern city, was a messy affair that demanded power plants be located away from homes. Generating electricity also demanded the economies of huge power plants. But today we can generate power efficiently at small scales, and integrate the means of generating power with the community and structures that need it. In so doing we reestablish ancient links between the city and its land, defined by light, water and wind.

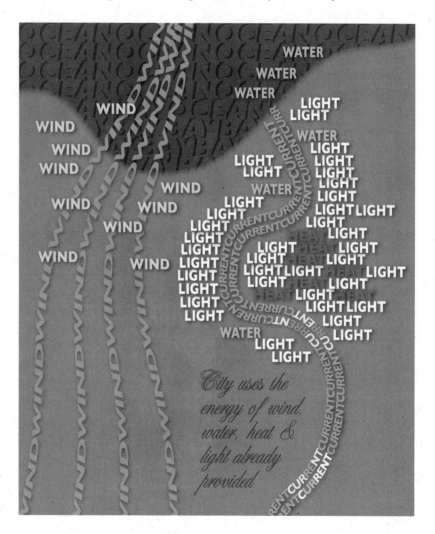

City uses the energy of wind, water, heat & light already provided

Schott Solar, BP Solar, Suntech, Motech and SolarWorld. Sharp is the world's largest producer with about a quarter of the market. Notably, China's production of PVs grew nearly 500 percent in 2004–05. The industry is developing in virtually all regions, with Japan being the leader in both private and public sector investments, as well as installed systems. In 1995 the global PV industry sold about 11,188 kilowatts of photovoltaic capacity, a measure of peak output in midday sun. Ten years later the figure was 301,530 kilowatts, a 27-fold increase.

Most PV sales are not in power plant form; they are small panels being purchased for homes, small-scale industrial uses and appliances, boats, recreation vehicles and village power. One 40-watt panel, costing a few hundred dollars, can supply enough electricity for a few light bulbs and a TV and/or refrigerator, and several can power a house. Some solar homes are in the middle of cities, but most are beyond the utility grid and rely on passive solar for heating, photovoltaics to generate electricity and batteries for power storage.

Millions of Tiny Power Plants

Photovoltaic cells, or PVs for short, are a type of semiconductor that receives light and generates electricity. These innocuous devices generate no noise, no smoke and have no moving parts, except photons and electrons. They are manufactured much like semiconductors, and are usually made of the same element — silicon. Tiny quantities of phosphorous and boron are also required in the silicon PV recipe.

Paradoxically, the very simplicity of photovoltaics prevents popular recognition of their existence. There are gas stations electrified by PVs, but the panels are just a roof over the pumps. There is a jail in California that's almost entirely powered by PVs, but few want to notice.

The vast majority of PV manufacture involves either single crystal or amorphous cells. Single crystal cells usually consist of round cells about the size of a coaster and set within a frame behind glass. Amorphous cells are little more than a thin film coating glass or stainless steel, and can be seen on a variety of small devices, such as wristwatches and calculators.

Unlike conventional centralized power plants, photovoltaics can be placed anywhere there is sunlight. In areas where there's insufficient sunlight the panels can often be placed nearby or augmented by other sources. Contrary to popular belief, PVs are viable all over the world, not just in sunny deserts. In areas prone to long gray days or weeks of snowstorms a structure may need a somewhat larger area of PV material, greater storage capacity and perhaps additional power from a community wind turbine or small hydroelectric unit in a river or tidal basin.

Photovoltaic cells reduce the entire process of generating electricity to a silent event happening in a space thinner than a human hair. This means photovoltaics can be one with the product needing the power, or they can

Photovoltaic cells are semiconductors. They can be made of silicon or several other materials. In manufacturing the cells the material is cast or deposited to form a thin crystal, then minute impurities are added, such as boron and phosphorous. In that moment a zone of static instantaneously forms — the "barrier zone." Above the zone there are a few too many electrons, and below there are too few, so as photons strike electrons the electrons bounce about seeking a new orbit with another atom. Finding none they have no other avenue except to traverse the wire, thus electrifying anything on it before finally reaching equilibrium in the lower half of the cell.

be integrated with other materials to create building products, thus solving two problems with one product. New forms of PVs are now being marketed as building products like shingles and plastic roofing that can be unrolled like tarpaper. There are PV walkway lights and streetlights, and all sorts of specialized PV power systems for navigation buoys, radio transmitters, billboards and other remote appliances. Designers have even begun experimenting with PV fabric, including electric camouflage military fatigues that integrate tiny PV cells and light emitting diodes so the fabric changes color to fit the surroundings — near invisibility. There are PV-powered unmanned

Photovoltaic cells are unlike all other forms of energy technology. "PVs" are emerging in new forms: a sandwich of lenses, photovoltaics and water tubes generating electricity and heat (left); a "cascade" cell of three layers, each tuned to different wavelengths of light, thus increasing efficiency to around 30 percent; windows of glass imprinted with photovoltaic film generating eight to ten watts a square foot; and possibly a TV screen where one layer of LEDs illuminates the image, while a second PV layer generates and stores electricity from light.

planes and even tiny PVs powering a radio transmitter on the back of a bumble bee unwittingly participating in a scientific study.

In the 1990s researchers envisioned photovoltaic films that could be printed on large machines, like magazine pages on a high-speed printing press. In 2006 several companies announced plans to do just that. One firm, Nanosolar, intends to build a factory capable of "printing" photovoltaics on a thin film much like printing the dots of a half-tone image on paper — only the dots are PV cells. The factory will produce enough photovoltaic film in one year to provide 430 megawatts, enough to power about 86,000 homes.

Unlike silicon photovoltaics, which utilize a crystalline structure spontaneously created when the silicon is formed, these new cells use compounds of metals and precision placement to create structures at a much tinier scale. If one were to look at the tiny dots on a printed page with an electron microscope one would see not flat dots but small mounds comprised of molecules, the structures of atoms representing the different elements required to produce red, yellow, blue and black. New photovoltaics are similar, only instead of molecular structures being placed to reflect a specific color they will be structures using very small quantities of certain metals and placed to respond to a specific color or wavelength of light.

Nanosolar intends to "print" copper-indium-gallium-diselenide structures using miniscule quantities of material per watt produced. Gallium is a soft white metal and a byproduct of aluminum production, while indium is also a soft metal, a byproduct of zinc production and as abundant as silver — Canada is the largest producer. Copper is also abundant.

Energy return on investment is a critical measure of efficiency. In calculating the return the key question is how long does it take for a new energy-generating system to produce enough energy to equal the energy consumed in building it? Any centralized power plant requires a variety of components and the construction of structures. The plant thus embodies the energy used to build it. It must operate for some time before its output equals this embodied energy; nuclear or fossil-fuel plants may not surpass the initial energy investment for 15 to 25 years.

Renewable-energy technologies in wind, solar-thermal and existing photovoltaics present a net energy profit within a few years. Nano-photovoltaics could provide an energy profit within a few *months*. Within months of the product's installation it is replacing energy that would have otherwise been produced by fossil or nuclear fuels.

Contemporary photovoltaics transform between 5 and 25 percent of the light striking the cells into electricity. In 2006 researchers working on new cell structures at the US National Renewable Energy Laboratory discovered how specific structures can result in photons doing what was thought impossible, causing not one but two to four electrons to move. The potential is a photovoltaic cell of 50 percent plus efficiency, or roughly 40 watts per square foot of electrical output at noon. That is the cutting edge of photovoltaic research.

Light is the fuel of the future. It was right in front of us all the time. It's moving right on in at the average daily rate of some 385 quadrillion megawatts just in the United States, a quantity far, far in excess of what we need. Light is everywhere and electricity can be anywhere.

5

From Light to Water

Inevitability of Renewable Energy

R ENEWABLE ENERGY IS FAVORED by national policy in most European nations and in much of the rest of the world, while in the US and Canada there is strong support in many cities and states, but less support expressed by the nation's energy policy.

The Bush Administration adopted a policy of force to control oil supplies and a return to nuclear power, offering only a modest commitment to renewable energy and a laudable but also modest program for hydrogen research. The administration's actions come atop efforts initiated in the early 1990s to create a free-market electrical grid, thus allowing utilities to sell and buy power nationally without regard for their historic territories. This newly opened market, coupled with pent-up demand, led to a boom in new coal and natural gas-fired power plants and efforts to restart the long moribund business of nuclear power.

These strategies will fail not because the technology doesn't work but because, compared to new energy technologies, the very notion of burning any fuel in some centralized facility and transmitting it great distances is archaic. Recognizing the simple facts that light and heat are widely distributed, and that contemporary conservation and renewable technologies are realities, leads to the stunning conclusion that we can not only end fossil-fuel consumption, but also develop a new energy system that does far more with far less of all resources.

Fossil fuels are over. Even with the best clean coal technologies, coal still involves massive machinery to mine, move and burn the coal. Using increasingly remote oil supplies, especially tar sands or oil shale, involves massive machinery to pump, transport and then burn the various fuels. Further development of oil, natural gas or coal resources represents scraping the bottom of the barrel, all to sustain outmoded technologies.

Oil and natural gas are rapidly diminishing in supply, with some experts suggesting we'll reach peak oil output globally around 2007; others say 2015 — it's a crap shoot. Peaking means the supply is declining steadily, and price is likely to rise steadily. Natural gas production is expected to peak and begin declining in roughly the same time period. We can resort to tar sands and oil shale, extracting crude at a higher cost and trashing landscapes at the same time, or we can grow grains for ethanol and almost certainly trash much of what's left of rainforests. These options merely sustain the unsustainable.

Transitioning away from oil is not merely a matter of switching fuel, but of transforming industry. Oil and natural gas are used in all manner of products, from tires to fertilizer to toys to computers to cars to rain gutters. A walk through any big box store is like walking through a catalog of oil products. Given this reliance, the rising price of oil means more than just a rise in gasoline prices. Crude-oil prices quickly show up in the price of plastics, which affects the price of innumerable products, and it doesn't matter if the product is made in North America, South America, Australia, Europe, Asia or Africa.

Paradoxically, the very nature of renewable energy prevents many people from grasping its extraordinary implications. Millions of people want to see a fuel like oil that can be burned, or some big machine that generates unlimited energy with no problems. They want a new energy factory to plug into the system we have. Seeing no such solution they presume there is no solution. They're looking for alternatives in all the wrong places.

We can go beyond debates about fossil and nuclear power and embrace a strategy that far exceeds the performance of existing technologies. It's a strategy that bypasses the whole complex of economic, political, military, scientific, health, technical and personal issues surrounding the acquisition, production and consumption of coal, gas, oil and uranium. It's a strategy based on the fundamental reality that it is possible to receive the energy already delivered by the sun and sustain human activities indefinitely with no reliance on fossil or nuclear fuels. It's based on the fundamental imperative that we have no choice but to develop renewable energy among all people. And it's based on the physical reality that mass production from a thousand

NO POLLUTION
DECENTRALIZED
CONSTANT PRICE
KNOWN EXPENSES
NO GRID REQUIRED
NO GUARD REQUIRED
RAPID DEVELOPMENT
AVAILABLE INSURANCE
WORKS FOR ALL PEOPLE
PERSONAL ENERGY SECURITY

SOLAR
BREAKTHROUGH

CENTRALIZED POWER
UNKNOWABLE PRICE
ENERGY INSECURITY UNAVAILABLE INSURANCE
NUCLEAR
HEAT POLLUTION REQUIRES GRID
UNCERTAIN DEVELOPMENT TIME
ENDLESS MILITARY GUARD
RADIOACTIVE WASTE FOREVER
NUCLEAR PROLIFERATION

Debates over nuclear versus solar-energy persist as if it were merely a technical discussion. But the realities that define the issue are not technical, rather they reflect the intrinsic nature of the two strategies. Nuclear energy, no matter how it's dressed up, carries with it risks by association with nuclear weaponry, and it requires an extensive fuel production and waste disposal system, a grid and centralized management authority. These issues translate to uncertain development time, unknowable costs and endless risk.

factories is more effective than specialized production from a few factories, and the political reality that individual initiative and local action from many points is infinitely more effective than executive action from a few.

The issue is simple. If we can reduce our energy consumption by more than half by using existing technologies, and generate ample electricity from our collective roofs, storing it in quantity, with no pollution at a reasonable cost now or in the near future, why waste another dollar or another minute doing anything else? We do not need to get energy; we're standing in it. Dollars from the sky and blowin' in the wind.

Efficiency of Wholes Not Parts

Critics of renewable energy typically make five erroneous assumptions. They assume electricity must be generated in a power plant; that existing generating facilities run 24 hours a day; that existing facilities are highly efficient; that renewable energy is insufficient to power big industries; and that price is all that matters. Based on these beliefs and, often, a dose of economic denial about the subsidies for conventional fuels, renewable energy is dismissed.

Solar energy is seen as weak, just as electric cars are seen as wimpy. But unlike conventional engines, electric motors can deliver full torque at any speed. A race between a conventional Grand Prix car and an electric car wouldn't be a contest. With no transmission and precise power control the electric car would lap the field. Similarly, few notice the extraordinary flexibility of solar technology. It can be used in dispersed form to heat water or generate electricity, and it can be concentrated to melt through steel. The sun generates enough energy to power us and all the plants we eat; all the trees our houses were made of; all the ships and animals that carried our ancestors; and all the wind, waves and clouds that define the lives of the trillions of organisms that populate earth — hardly a wimpy performance.

Renewable energy is often criticized as being inefficient, a characterization that adds to the wimp factor, with the implication that existing technologies are very efficient. Citing the low efficiency of photovoltaics in converting to electricity a seemingly meager 12 percent of the light that strikes the cells, critics dismiss the technology. They compare such low efficiencies with that of a big power plant, where more than 50 percent of the fuel consumed might be transformed into useful electricity. This is a dubious comparison.

Efficiency is a measure of economy. A machine is said to be efficient to the degree it uses the least resources to accomplish the most work. But we often look at machines in isolation. Much like assuming cars are cheap without counting the cost of roads, proponents of one or another technology often claim their concept is very efficient at generating electricity, but neglect to note the energy lost in the whole system of distributing electricity.

We build the power plants...

to generate the electricity...

losing half in transmitting it...

to the buildings of customers...

who lose half.

ELECTRICAL CAPACITY
POWER PLANTS OF ALL KINDS, ASSUMING ABOUT 40 PERCENT ARE SHUT FOR MAINTENANCE AT ANY TIME

ELECTRICITY GENERATED
OUTPUT OF OPERATING POWER PLANTS AT ANY TIME

ELECTRICITY WASTED
OUTPUT LOST IN TRANSMISSION

ELECTRICITY CONSUMED
OUTPUT TO USERS

ELECTRICITY WASTED
AT HOME

ELECTRICITY NEEDED
HOME

The electric-utility system in North America and much of the world is based on largely centralized power plants transmitting power via long-distance powerlines and local distribution grids. On average the distribution system loses about 60 percent of the power in transmission. Then consumers waste about half of what they receive in low-efficiency structures and appliances. Industrial and commercial customers waste less, but often maintain old and inefficient facilities. Aged distribution systems in many countries probably lose more than 75 percent.

Coal under the high plains of Wyoming is potential electricity. But first it needs to be mined, and that takes energy in the form of big shovels. Second, it needs to be moved, and that takes the energy of many trains. Third, it needs to be burned in power plants, and that loses energy in wasted heat. Fourth, the electricity needs to be transmitted to consumers via the grid, and that loses energy by the resistance in wires. In the end, after subtracting all the energy consumed, it's likely that only about a third of the original potential energy in Wyoming actually arrives at one's toaster in the kitchen in Chicago. On average the means of producing and distributing electricity consume about two-thirds of the energy.

If the majority of homes converted to independent or neighborhood-scaled local-grid systems, with hydrogen and/or batteries used for storage, the sun could power all structures. The amount generated, primarily by photovoltaics plus wind turbines in some areas, would range from 2.5 to 3.5 times maximum consumption on any day, and could recharge hydrogen tanks and/or battery packs for night and cloudy periods. Assuming new technologies are used, the future home would use less electricity than it does today, thus reducing the size and cost of the energy equipment, and representing less total generating capacity.

yet requiring a fourth the investment in generating capacity, or roughly equal to the total electricity now consumed.

& use sunlight to power buildings, producing up to three times the power needed to cover storage needs...

We can use new tech to cut consumption two-thirds...

ELECTRICITY HOME NEEDED

ELECTRICITY HOME NEEDED

ELECTRICITY ON OR NEAR HOME GENERATED

ELECTRICITY CONSUMPTION TODAY CONSUMED

ELECTRICITY BY CONVENTIONAL UTILITY SYSTEMS TODAY GENERATED

ELECTRICAL INSTALLED CAPACITY OF ALL POWER PLANTS CAPACITY

Which is more efficient, a system that consumes more than half the energy it generates in machinery stretched out over the landscape, or a system that consumes a tiny quantity of steel and silicon to produce electricity from a roof only feet from the toaster? Does it matter if photovoltaics are only 10 or 15 percent efficient if a typical home can be powered with PVs covering half the roof? New photovoltaics achieve 18 percent efficiency, and others in development prove efficiencies exceeding 25 percent. There's no shortage of sunlight.

Presuming photovoltaics must be concentrated in a power plant, critics envision vast remote fields devoted to PVs. They succeed in transforming a solution into a problem.

The notion that electricity is made in a factory is couched in a 19th Century vision of power generation, where noisy smoke-belching power plants were far away on the other side of the tracks. Renewable energy technologies, especially photovoltaics, don't need to be isolated because they just generate electricity, nothing else. The very qualities that make them valuable make them unnoticeable.

Roofs are an under-utilized resource covering more than half the area of cities, and they are right over our heads. The potential of electric roofs is obvious from the window on any flight approaching any airport into any city. Roofs dominate the city landscape. The vast majority of buildings are one-to three-story homes, apartment buildings, retail complexes and industrial structures. Photovoltaic roofs, with solar-thermal systems on the roofs of larger industrial and office buildings, can provide ample power for the vast majority of structures.

Roofs can be more than just shelter. The flat urban roofs of many homes could be part photovoltaics and part grass, with the latter creating a cool roof with perhaps a deck. Instead of being just an expense, the roof becomes a means of keeping warm, generating electricity, collecting rainwater, oxygenating the air, and perhaps visiting with friends. This is maximizing your asset value.

Renewable energy is as practical for powering a factory as a watch. Judging by the sheer size of many industrial and commercial structures it would seem they consume a vast quantity of electricity. Yet San Francisco's Moscone Center, a massive two-block-square underground convention center, receives half its electricity from photovoltaics covering a quarter of its roof. Where higher power densities are required, roof-mounted solar-thermal technologies can supply two to four times the energy in the same area. The strategies would vary by building, but in the vast majority of cases all needs could be met on or near the structure.

Solar-thermal power generators are ideal for producing heat *and* electricity. Contemporary systems use mirrors to focus sunlight on pipes carrying oil and are generally being built by utilities as large power plants. The hot

It is often said conventional utility systems are efficient and photovoltaics are not. This is a questionable assumption. Electric-utility systems are built around centralized power plants transmitting power via grids and consuming a fuel. Obtaining the fuel, processing it, moving it, burning it and distributing the product, electricity, can consume from 75 to 90 percent of the original energy value of the fuel. Thus the entire system is only about as efficient as typical photovoltaic cells, which can be located where the power is needed and use fuel delivered daily — light.

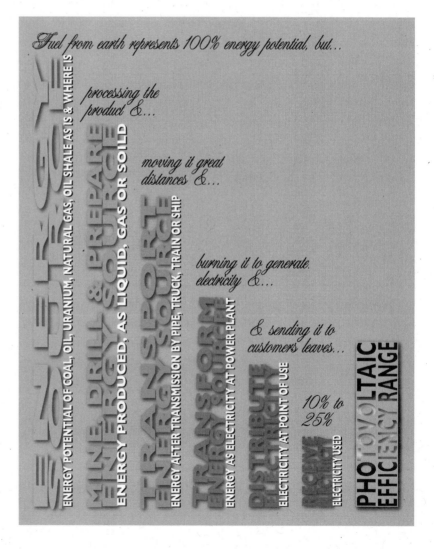

oil is continuously circulated through a boiler to generate steam, which is used to spin a turbine to drive an electric generator. The same system can heat a space or use hot water for an industrial process. Solar-thermal generators could be mass-produced in several sizes, from a small plant the size of a standard gasoline generator to larger units designed to be installed atop factory or office building roofs or over parking lots.

There is a specific type of solar thermal collector, called a non-imaging concentrator, that exemplifies the simplicity of renewable technologies. Its purpose is to concentrate sunlight to achieve higher temperatures. To do that it's been assumed that the device must be centered on the sun constantly; thus a small guidance system, analogous to celestial drives that move a telescope to follow a star, would be required to track the sun by day and season. By contrast a non-imaging concentrator is a mirror shaped to a precisely calculated curve of changing radius, much like the inside of a small teacup, and because sunlight striking that particular curve will be reflected downward, no matter what time of year or day, no tracking system is required. Non-imaging concentrators can be cone-shaped or long troughs aligned east to west, but in either form they need only be open to the sunlight. Light striking interior walls will be reflected like water spiraling down a drain. A small non-imaging concentrator, say a metal cone about one foot tall, with the interior wall coated with chrome, can focus about 860 watts on a ceramic disc half the size of a thumbnail. It would be white hot.

Photovoltaics are viable for the majority of residential, small commercial and stand-alone purposes, while photovoltaics plus solar thermal, wind or small-scale hydroelectric power would be required for larger industrial applications and major complexes of buildings. Solar-thermal facilities are the most attractive option for large-scale structures and industrial complexes because they concentrate so much energy in a small area, with the capability of generating both electricity as well as heat as hot water or steam.

Bottling Light and Water

Critics claim renewable energy isn't practical because the sun doesn't shine 24 hours a day so a big power storage system is needed. They compare this intermittent performance with existing electrical generating stations, which are assumed to operate 24 hours a day. In fact, all power plants, whether they burn coal, oil, gas or uranium, are shut down for maintenance as much as half their average working life of 40 to 50 years. Utility companies regularly turn one plant off and another on, so everyone thinks the whole system is working 24 hours a day. The need to operate multiple power plants is equivalent to the need to generate and store an excess of renewable energy when the sun is shining or wind blowing.

Photovoltaics and solar-thermal technologies are intermittent on a daily basis, losing output when it's overcast, but they keep right on generating power day after day, year after year. Wind turbines are even more intermit-

Existing electrical systems do not store power, as one might store gasoline in a tank, rather they are on 24 hours a day. Given weather and darkness it's imperative renewable energy be stored. There are essentially two methods, hydrogen and batteries. The hydrogen-fuel-cell process, using water as the hydrogen source, is more complex than batteries, but it offers pure water and heat. Unlike batteries, which degrade over time and ultimately require replacement, a hydrogen system represents stable energy storage, plus quick recharging of vehicles.

ENERGY STORAGE

TECHNOLOGY	PROCESS		PROCESS	TECHNOLOGY
SUN	LIGHT & HEAT		LIGHT & HEAT	SUN
PHOTOVOLTAIC WIND TURBINE SOLAR-THERMAL WATER TURBINE	ELECTRICITY GENERATED		ELECTRICITY GENERATED	PHOTOVOLTAIC WIND TURBINE SOLAR-THERMAL WATER TURBINE
ELECTROLYZER	ELECTRICITY CRACKS WATER INTO HYDROGEN AND OXYGEN		ELECTRICITY STORED IN BATTERY	NICKEL HYDRIDE LITHIUM-ION LEAD ACID NEW TECH
HIGH PRESSURE METAL HYDRIDE	HYDROGEN STORED IN TANK OXYGEN VENTED TO AIR			
	HYDROGEN PLUS OXYGEN FROM AIR			
FUEL CELL	GENERATES ELECTRICITY AND HEAT AND		WITHDRAW ELECTRICITY	
WATER TANK	PURE WATER			
APPLIANCES VEHICLES BUILDINGS VEHICLES GLASS	ELECTRICITY POWERS HEAT WARMS WATER NOURISHES		ELECTRICITY POWERS	APPLIANCES VEHICLES

tent, but they too are very dependable and will be spinning for decades. Photovoltaics are so simple there just isn't anything to break. Occasionally dead leaves and dust need to be washed off.

Batteries store solar-energy. Contemporary deep-cycle batteries, combined with photovoltaic panels, now power millions of navigation buoys, telephone relay stations and other vital communications links. Photovoltaics and deep cycle batteries are used for these facilities, which cannot fail because lives depend on them, precisely because these technologies are so dependable and require so little maintenance even in the most hostile environments. Thousands of solar homes rely on batteries, usually filling a small closet, to keep lights and other appliances going even through storms lasting a week or more. Thousands of RVs and boats use batteries with PV recharging.

The newest batteries, notably Ovonic batteries developed by Energy Conversion Devices and Chevron-Texaco, can store even more power in less space, and can be recharged in much less time. Lithium-ion batteries are similarly practical and are used in consumer electronics, although they do degrade and require replacement.

Hydrogen can be used to store solar-energy. Electricity, generated by photovoltaic, solar-thermal, wind or hydro facilities, can power all sorts of appliances in a building, including an electrolyzer, a device that uses electricity to divide water into its constituents, hydrogen and oxygen. The oxygen is then released to the atmosphere where it disperses, while the hydrogen is stored in a tank. Later, as needed, the hydrogen is drawn down to supply a fuel-cell, a kind of reverse electrolyzer, which re-unites the hydrogen with oxygen drawn from the atmosphere, in the process generating electricity, heat and pure water simultaneously.

Compared to batteries hydrogen is more portable and the process of using it results in not just electricity on demand, but useful heat and pure water too. Hydrogen can be stashed in tanks of any size, both fixed and portable, or piped much like its hydrocarbon relative, natural gas. Using hydrogen as a storage medium is equivalent to storing or moving energy and pure water simultaneously.

We live by a star made largely of hydrogen. Hydrogen is one of the most plentiful elements in the universe. We are primarily made of water with various minerals in solution, and since water is two parts hydrogen we are more than half made of hydrogen, with a large helping of oxygen. Gasoline, diesel fuel, natural gas and coal are all considered hydrocarbons because they are primarily hydrogen and carbon. Virtually all air pollution is a result of burning hydrocarbons — wood or fossil fuels. The hydrogen burns in an engine or power plant, leaving behind the carbon in the exhaust. Most pollutants are removed, but not the carbon dioxide, nitrogen and water vapor.

Hydrogen burns. When the space shuttle lifts off, its main engines are receiving hydrogen and oxygen. The two gases arrive in the rocket engine and in milliseconds go from super cold liquids to a explosive fire to a white plume of steam as the rocket goes from zero to thousands of miles per hour. The secondary boosters use solid fuel made of aluminum powder and an oxidizer.

Fear of hydrogen was established on a cloudy day in 1937 when the hydrogen-filled German dirigible "Hindenburg" approached its landing mast in Lakehurst, New Jersey. According to one of the most plausible explanations static electricity caused the airship's skin to ignite. Unbeknownst to the airship's owners the silver paint recipe included oil and aluminum powder — now known as rocket fuel. The flame traveled along the skin in a flash, igniting the hydrogen bags to form a huge ball of fire. The dirigible was gone in seconds. Although 61 of the 96 passengers and crew survived the event the horrific images convinced millions of people hydrogen was far more dangerous than it really is.

Hydrogen is no more or less dangerous than gasoline or natural gas. Spontaneous combustion of any flammable substance is not possible without air and a source of ignition. Inside a gasoline tank two elements are present: gasoline fumes and air. Inside a hydrogen tank there's just hydrogen. Moreover, if a hydrogen tank leaks it doesn't flow like water nor hug the ground like natural gas. It rises and disperses, and unless ignited at a specific moment it simply dissipates. Gasoline can explode and flow like water — a burning river. Natural gas can fill a building until a source of ignition touches it off. Managing hydrogen is no more or less difficult than managing natural gas.

Leakage is a problem given hydrogen's propensity to sneak through just about any opening. Excessive free hydrogen could exacerbate climate change. But then hydrogen will also be costly and would probably be more expensive than gasoline, so users and manufacturers will have an incentive to maintain systems that don't leak. Tight hydrogen storage tanks, pumps and piping exist and the technology of building such systems is established.

Today energy companies intend to use natural gas as the hydrogen source. This strategy builds upon the existing hydrogen industry that now produces hydrogen for various products, such as hydrogenated vegetable oil. The technology involves separating hydrogen from carbon and stashing the carbon somewhere, such as pumping it into deep rock strata, while hydrogen is distributed via truck to service stations.

It is probable, not certain, that tomorrow's energy companies will grasp the fundamental reality that renewable energy technologies, along with hydrogen storage systems, can be as decentralized as the sources and uses of energy, and that this strategy is more efficient and reliable than centralized facilities and marketable anywhere. Electricity and hydrogen can be generated where needed, so big power plants and related facilities to produce hydrogen, as well as massive pipeline or trucking delivery systems, are simply unnecessary. These technologies change our criteria of what is possible by making the *means* of generating and storing energy available to all at all scales.

Price is Not the Only Value

It's widely perceived that any new energy source will not succeed unless it's cheaper than the cheapest existing source. But utility prices are in part a function of region and vary from 3 to over 12 cents per kilowatt hour. In most of the US and the world, wind turbines are price competitive now as are solar-thermal facilities that concentrate solar heat and convert it to electricity. PV electricity is priced at about 25 cents per kilowatt hour in 2006, but with tax credits it could be significantly less. In many markets photovoltaic home systems are competitive due to tax credits and utility interconnect systems. Many homeowners are experiencing *lower* utility bills because they pump more power back into the system than they consume.

PV technology is often price competitive today within and beyond utility grids. Beyond the grid, which includes millions of square miles of semi-wilderness ringing thousands of small towns, PV electricity is often less costly than the cost of extending a utility line. Within grid territory PV systems can be very economical now if tied to the grid. The $5,000 to $15,000 investment can be offset by other values; since PVs are also functioning as roofing, about 30 percent of the cost is a roofing expense. If they generate power to the utility the credit reduces their bill.

Fuel-cells today are analogous to where photovoltaics were in the 1990s — too expensive. The technology has attracted many innovators because they all see a technology with global potential. They also see how hydrogen fuel-cells represent new values, not just of nonpolluting electricity but of pure water and heat. Given these exceptional values there is reason to believe the technical problems will be addressed and costs will decline.

Seeing today's prices as the prices of tomorrow ignores inevitable technological changes, market forces and the unique nature of renewable energy. Wind turbines, photovoltaics and other renewable technologies, as well as fuel-cells and hydrogen systems, are all fixed devices that use the "fuel" of light, wind and flowing water. If integrated with buildings these devices would be financed as part of the building. Thus the cost of energy becomes a fixed cost included in the mortgage or lease payment, a dependable budgetary item. Meanwhile fossil fuel prices may go up or down with the vagaries of war, weather and politics, but in the foreseeable future they will almost certainly trend up. In any case the price and uncertainty would be meaningless to those relying on renewable energy.

Price alone rarely determines the success of new technologies; value does. Trains, light bulbs, cars, radios, computers and innumerable other innovations were initially expensive, and although prices declined through mass production, few of these technologies were ever cheaper than those which they replaced. Installing a lighting system was not cheaper than kerosene lamps. A century after cars were invented they remain — to buy, operate and maintain — at 60 cents per person per mile more than twice as expensive as trains, which average 25 cents per person per mile. If all hidden expenses are included, such as free parking and the uninsured cost of accidents, cars would be even more costly. Yet cars offer some values unattainable by trains, so the extra cost has been absorbed in government subsidies, health care expenses and personal budgets, such as the cost of a two-car garage.

Analogous to cars, and unlike oil, natural gas, nuclear and coal, renewable technologies, especially photovoltaics and hydrogen, represent new values impossible to achieve using any of these older technologies, specifically the ability to use one mode of energy technology for structures, industry and transportation, and to develop it as decentralized facilities that incorporate water purification and offer un-interruptible power. These new technologies simply dispense with whole categories of other impacts, such as pollution, reliance on distant energy sources and the damage to land and ecosystems caused by mining, drilling and radioactive waste. If we define an economy as a system of values represented by money, then it could be said new technologies often establish a new economy.

Despite the nuclear industry's claims of a good safety record, a claim made because power plant workers aren't dying in accidents, it would be absurd to assume nuclear power is safe. The paradox: reactors can be safe to operate, but they are all vulnerable to catastrophic failure with horrific consequences. Chernobyl's reactor didn't just melt down, it contaminated a vast area of land now poisoned for generations. The reactor at Three Mile Island near Philadelphia did not melt down, but came so close to it that the governor of Pennsylvania debated whether to evacuate the entire state, knowing that many residents might never be able to return. Given these liabilities, and the issues of unresolved long-term waste storage, nuclear proliferation and terrorism, private insurers will not insure nuclear power plants. These other liabilities, and the costs of coping with existing power plants soon to be retired, are already in *everyone's* budget as government expenditures.

In relation to climate change nuclear power is often cited as the ideal because it produces no carbon dioxide, methane or nitrogen. New "pebble-bed" reactors are designed to be inherently safer and could be mass-produced. But building such plants would consume considerable quantities of fossil fuels — notably oil in construction. Supplanting all energy production with nuclear plants, for powering buildings, industries and transportation, represents a long-term construction effort of staggering magnitude.

If we value energy technology honestly and include all costs, renewable energy is economically viable now. There's no hidden waste problem, no risk of catastrophic events, no noise, smoke or other environmental impact, no problem buying insurance and no need to fear the loss of fuel to terrorists.

Renewable energy already uses photovoltaics, with car batteries for storage, to power phones, televisions, computers, small refrigerators, lights and electric hand tools; it can set-up shop anywhere, with no noise, no wires, no smoke and no trace when it leaves.

Dismissing renewable energy as too expensive is a denial of the subsidies granted other forms of energy and of the extraordinary new values it embodies. What's the value of an energy source that allows buildings to be powered independently, or via local-area grids, connects a complex of buildings, and thus eliminates the risk of blackouts? What's the value of a product that ends reliance on fossil and nuclear fuels and offers not just national energy independence, but individual independence? What's the value of an energy source where commodity prices don't fluctuate, and where consumers pay a fixed monthly cost? What's the value of technologies that would make energy so plentiful no one would care how much one's car or home consumes?

What's the value of a city capable of sustaining itself? San Franciscans, including their daily guests from the suburbs, consume 17 million kilowatt hours of electricity in a day. The city covers 49 square miles, with around half of that area consisting of roofs. Although the city is legendary for its fog it is in fact bathed in sunshine most of the year. Besides, fog only reduces photovoltaic output; it does not stop it. If our objective is to make the city independent of external energy, thus generating sufficient power on most days to cover its immediate needs and provide storage for night and cloudy periods, the photovoltaic cells would cover around 20 percent percent of the city's area — mostly as roofing. Generally this means either photovoltaic roofing or panels atop the existing roof would cover half to three-quarters of the average home roof, but if electrical consumption is cut 50 percent by using high-efficiency appliances most homes could provide sufficient power for one or more vehicles as well. Throw in some tidal power under the famed Golden Gate Bridge, plus windmills along the windy coast and atop many high-rises, and the city would be independently powered and invulnerable to blackouts, thus safer in the event of a major earthquake. In addition, the entire system is simpler and less costly than any form of centralized power, and no one could do an end run around the city.

Why go to all the trouble and expense of sustaining power plants, electrical grids, offshore drilling rigs and global supply chains of wires, tankers and pipelines if we can obtain ample power from a film thinner than paint that can be integrated with the roofing, wall panels and even windows of the building that needs the power? Why maintain all that technology of gathering fuels, generating heat and distributing energy if we have ample energy right where we are?

6

Distribution by Rain

A Question of Purity?

WATER ISN'T MANUFACTURED in reservoirs and isn't a product. Seeing water as a product is a denial of the nature of water. If we named every living thing by its primary constituent we'd all be in the Water family living on Water Street in Watertown in the Global States of Water. Water isn't part of the commons: it is the commons. Today "I" am the water recently of Tahiti, and we are the water that once called itself Roman, Mayan or Chinese. We've all been flowing along for quite awhile now.

Drugs — legal drugs — are among the most vexatious pollutants contaminating this commons. Many components of synthetic drugs, carried in human waste, do not break down in sewage treatment plants and are thus passed into the water. Hormones in women's birth control pills, for instance, are not entirely absorbed by a woman's body. A large portion ends up in the water. Antibiotics and a variety of other medical substances are also in the waters, often causing problems in the reproductive cycles of wild species.

Water is not apart from us; rather, it flows through us, both as liquids we drink and as moisture our skin absorbs or exhausts. Thus we are all but contiguous with river, sea, rain, snow and ice, distinct only by the relative measure of water in us and moisture in the atmosphere around us. Polluted water generally can pass into us from rain, drinking water or from bath water.

Today the average American carries various pollutants in their bones and tissue. It doesn't matter whether one is rich or poor, we are all polluted with industrial chemicals and drugs. We are left with a legacy of uncertainty, an endless range of possibilities as to which substances may cause what malady, from cancer to birth defects. No amount of study can ever determine all the consequences of poisoning all of us. Polluting the waters is a gross violation of the public trust.

Given what is known about chemical and biological pollutants, about

Water is not apart from us, nor is electricity. Water and electricity flow through, on and around every plant and animal living on or in the earth. We are mostly water, plants are mostly water and water permeates the air and the earth. Since water conducts electricity the electrons buzzing about the planet tend to follow water vapor, water columns and underground waters. Electricity and water are bound in a continual dance, with our very thoughts defined in part by the pathways of electrons traveling within the sea of cells in water that is our brain.

our vulnerability to terrorism via water and about the primal value of water, it would seem imperative that we establish a new standard of water purity. Water cannot be just be another product in the economy because water is the currency of all economies. There is no such thing as private water. It's all common.

Purity must be the standard by which all strategies and technologies are measured. The value of purity must become inviolate. Equivocating about purity, and allowing anyone to poison water is equivalent to allowing

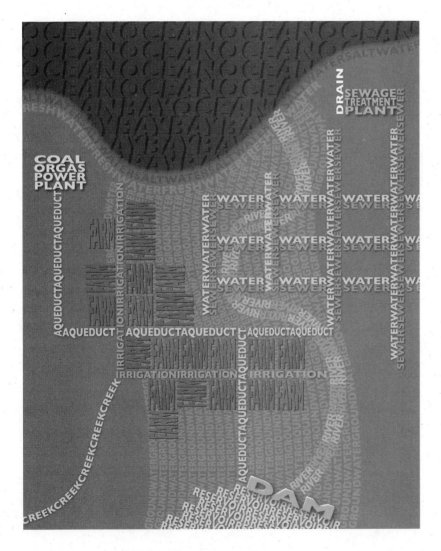

When most contemporary water systems were built there was little understanding of the workings of rivers and groundwater, so dams were built to store water while the largest reservoir of all, the earth, was ignored or pumped out. In many coastal regions groundwater has been reduced, usually by irrigation, to where wells are pumping saltwater. Most cities have two water systems, one to bring in freshwater and a second to remove wastewater, while a few also have extensive storm drain systems. Cities with tertiary treatment plants — most North American cities — virtually purify the water only to drain it.

anyone to injure or kill others with no consequences. Mr. and Mrs. Water didn't vote to pollute themselves.

Roman Aqueducts to American Spas

Water is increasingly viewed by many experts as the next source of wars and conflicts, presumably after we finish fighting over oil. It would seem, given that many cities in industrialized nations have already exhausted available local water sources, and that cities in less developed nations are rapidly exhausting available supplies, that water will indeed be a scarce resource spawning conflicts.

This view of water as a diminishing resource is based on the pervasive belief in scarcity, which is in turn based on a simplistic view of what water is, where it comes from and how people use it. We all drink anywhere from one to four gallons of water a day, often in the form of other beverages, but those in industrialized countries may consume several times that amount in toilets, washing machines, showers and gardens — up to 150 gallons per person. This flow of water is widely perceived as one-directional: from source to drain. But since few people know the location of their particular source, or the ultimate destination of the drain pipe in their bathroom, they act as if it came from nowhere and goes nowhere. Water is ignored once pissed into sewers. It's used water.

Water today is managed much as it was in Rome, as if nothing had changed in 2,000 years. Until the last two centuries little *had* changed, as all civilizations merely copied the Roman model, except with pipes and underground sewers instead of open aqueducts and gutters. We did deal with the stench in the streets.

Again and again cities formed around bays and rivers. As citizens exhausted wells and polluted creeks, they built reservoirs to get pure water from upstream. As the cities grew they needed more water, so they built reservoirs further upstream. Meanwhile the citizens kept draining polluted water into the nearest watercourse, lengthening the drainage pipe so on windy days the smell wouldn't remind everyone of where the water went.

Everyone was so focused on pure water from far away that almost no one noticed how acres of impermeable roofing, plus pavement carpeting up to 60 percent of the land of many modern cities, were acting as a vast rain collector. Increasingly cities had to construct larger sewers or separate storm drains to cope with rain.

Inevitably, as cities grew the concentration of pollutants overwhelmed their watersheds' ability to digest it. The combination of raw sewage, storm drains and industrial chemicals generated an often smelly stew. By the 1950s it was obvious this could not continue, and the resulting political pressure, at least in most major cities and developed nations, resulted in the passage of anti-pollution laws and construction of sewage treatment plants that at least treated the water. Since then tertiary treatment plants have been built in many cities, and these facilities virtually purify water — 99 percent plus.

Meanwhile, agriculture was becoming agribusiness, a potent political force by the mid-20th Century. The dominant farming paradigm was one of industrial production, as if a field were an assembly line, soil the raw material and water just a production expense. Sustaining production was more important than sustaining soil or the water table. Inevitably groundwater declined and getting more water became imperative.

Eighty years ago, society's understanding of the natural flows of water and its relationship to land was heavily imbued with an industrial vision of the natural world. Floods were seen as Mother Nature on a rampage, hardly a natural process of restoring soil and recharging the water table. Groundwater was invisible and reservoirs very visible. Cities needed pure water, dams could be built on a huge scale and besides, they'd also generate clean hydroelectric power. It seemed to be a stroke of brilliance, all those values from one gargantuan chunk of concrete.

A Change at the Water Factory

The vast majority of the world's citizens in industrialized countries view reservoirs and aqueducts as analogous to power plants and electrical distribution systems, with both seen as industrial facilities that produce water or energy. This mechanistic view tends to place emphasis on the visible reservoir and the rivers and creeks that feed it, not the invisible quantity of water evaporating from the reservoir, nor the underground rivers and aquifers flowing just beneath one's feet, nor even the flood of stormwater that drains off the impermeable walls, roofs and roads of cities.

Contemporary water systems are a world of paradox. We now rely on systems where reservoirs can lose more water to evaporation than they distribute to customers; where extensive aqueducts supply pure water to cities where many residents prefer bottled water; where cities develop elaborate storm drains to accommodate rainwater that often exceeds what they receive from reservoirs; and where pure water from far away is dirtied, then virtually purified, only to be thrown away.

However, this view of water is changing. Southern California water agencies are increasingly recognizing that the earth, in the form of a water table, is potentially a far more efficient means of storing water than any

Cities could shift to local waters. If new energy technologies are fully utilized it would be possible to eliminate power plants, dams and centralized sewage treatment plants. Cities would rely on groundwater, using wells to pump water up or rainwater down, thus replenishing recycled water and storing rainwater for landscaping. Sewage would be piped to neighborhood facilities in parks — an excuse for fountains and ponds every few blocks — or to a restored local marsh. Farming, due to increasingly chaotic weather, is likely to become more diverse. Restored groundwater would force saltwater back to the ocean.

Cities of the future can rely on water & sewage systems that utilize natural flows of water

reservoir. Though it does lose water downstream to a river or the ocean, it doesn't lose water to evaporation because the water isn't exposed to the sun. The water table also functions as a natural filter. Best of all it's right under the city.

Over the last decade, local agencies have been expanding their reliance upon groundwater by developing catch basins, essentially gently sloped park-like areas. Catch basins function like seasonal wetlands that flood after a heavy rain. Over the days following a rain much of the water seeps into the earth. In some cases wells inject water to accelerate the process. Los Angeles now receives 15 percent of its water from right beneath the city. The city is also pumping water from sewage treatment plants back into the water table.

Some Good Shit

Restoration of groundwater is also facilitated by new sewage-treatment strategies. Conventional sewage treatment emphasizes biological digestion and chemical additives to purify water. Building on contemporary understanding of ecology some water agencies and private companies are developing wholly biological methods that employ not only microorganisms to break down the sewage, but also wild plants to convert the waste to biomass, supporting the growth of fish and flowers. These techniques mimic the ecology of marshes on a small scale.

Biological sewage treatment can involve real marshes. Just south of the small town of Arcata, California, there is an extensive marsh where visitors can walk the pathways that lace the town's wild sewage ponds and see fish amidst the reeds, as well as ducks, egrets and hawks. The place is alive with life, much of it spreading into the reeds and salt grasses along the bayshore and transforming former lumber mill sites into habitat.

It is possible to develop biological marshes in the midst of big cities. Even individual buildings or building complexes can feature a series of ponds, with water gently trickling from one pond to the next amidst a forest of marsh grasses, all fertilized by the sewage from the building, yet leaving only the odor of flowers.

Biological sewage-treatment facilities exemplify a larger trend, and a larger vision, of developing water systems that work like nature. Existing systems are imbued with the belief we could craft more efficient mechanical systems. Between the mid-19th and late-20th Centuries we transformed entire watersheds into vast water and sewage treatment systems. These magnificent creations, such as New York's pioneering Hudson River system and the huge California Water Project, seemed testament to our wisdom. Unfortunately these systems didn't include the tiny creatures working the marsh, the global rivers of fish and birds, and the steady percolation of floodwaters with salts to the sea.

A city utilizing rain as its primary source would use streets and roofs to capture water, pumping much of it underground. Wells would provide water for buildings, gardens and firefighting. Sewage treatment would be scattered around the city, and usually incorporated in new parks, from "pocket parks" to large public gardens offering fresh produce and flowers. All parkland, medians and riverside lands would be vital for absorbing rainwater.

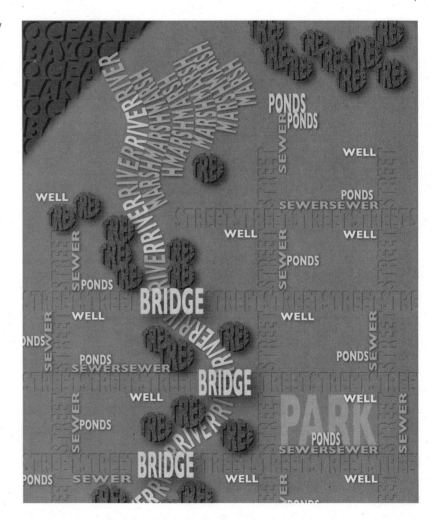

7

One Water and Energy System

From Many Technologies to One

A LEAF IS A MULTITASKING DEVICE. It receives and converts solar energy into useful chlorophyll, which the tree's cells then transport to sustain the tree's growth. It captures and directs rainwater to the tree's roots. It shades the ground from excessive heat that would dry the soil. It provides food for insects that feed on parasites that would otherwise kill the tree.

Today it's assumed energy and water must be provided by several separate systems each doing only one thing. In most cities we now rely on one water system for incoming water from reservoirs, a second water system to drain sewage and in some cities a third water system to handle storm runoff. For energy we rely on one system distributing electricity, and often a second for natural gas, fuel oil or propane. Vehicles rely on separate gasoline and diesel fuel systems. It's assumed we must continue to rely on multiple systems because each technology can only do one thing.

New infrastructure technologies suggest the potential of an integrated energy and water system that functions like a natural system. Photovoltaic roofs would, much like leaves, generate energy from light and collect water. Using a slightly more sophisticated version of conventional downspouts and rain gutters, routed to cisterns and wells, the water would be collected, stored and often pumped underground. The electricity they generate would crack some of the rainwater they collect, storing it as hydrogen for future use. When the hydrogen was recombined with oxygen in a fuel-cell it would produce electricity and pure water. Such a system could provide us with clean energy and pure water.

If fully developed this new infrastructure would translate to major reductions in urban infrastructure. Homes would recycle water and rely on the water table beneath for additional water. Most would depend on photo-

The proposed energy-water unit would perform many functions. It would receive electricity, powering an electrolyzer to crack water into hydrogen and oxygen. In recombining the gases the fuel-cell would generate electricity, pure water and heat. Its air cleaner would remove pollutants that might harm the fuel-cell, as well as tiny quantities of carbon dioxide and methane, which would be periodically removed as waste cartridges for sequestering.

voltaics for power. Streets would no longer have water mains, underground gas or electrical conduits, or power poles. Clusters of homes or commercial structures would share a local-area grid of hydrogen pipes to balance the energy inputs and outputs of several structures. In higher density suburbs sewer systems would be smaller and direct sewage to a new neighborhood park for transformation into flowers, fertilizer, fish meal and fresh water. Rainwater would collect in the streets and drain into parking areas with permeable pavement where grass would grow. There would be one local sewer pipe and one fiber-optic cable down the street.

Energy and Water Daily

Imagine a typical block of 12 one- and two-story ranch-style homes in a typical suburb of a typical city in the year 2020. Six homes might have hot tubs, and all share a common swimming pool. The homes were built between 1970 and 1980 and still look much the same, except the roofs are shiny and the backyards are divided into private spaces and one common space with the pool and garden.

The photovoltaic roofs would glisten like faceted diamonds. Every sunny day the roofs would generate electricity. Electricity would power a water pump beside the pool and electrolyzers in each house. Electrolyzers, a long-established technology, would crack water into hydrogen and oxygen. The oxygen would be vented to the atmosphere while the hydrogen is stored in a tank under each garage. On sunny days the roof would generate electricity and on rainy days it would catch water. Water funneled by rain gutters and downspouts would flow into home water tanks, hot tubs, pools or cisterns at intersections, with overflow pumped into a well, or drained to a restored local creek.

As homeowners turn on lights and appliances hydrogen would be admitted to the fuel-cell, where it would recombine with oxygen drawn from the atmosphere to generate electricity and pure water. Pure water would be stored in a glass-lined tank for drinking, bathing and cooking. The two water tanks, as well as hydrogen tank, pumps, electrolyzer and fuel-cell, would occupy about the same space as a conventional water heater and furnace. Additional water, supplying the pool, hot tubs and landscaping would be available from wells — one per block.

Each house would include new appliances. Showers would have one valve with a temperature indicator, from cold to hot by degree, and sink faucets could supply water for tea or coffee — the faucet would be the heater. Toilets and washing machines would incorporate water tanks and filtration systems. Homes and commercial buildings would be illuminated more by sunlight during the day, using fiber-optic lighting fixtures that automatically turn up electric lights when sunlight diminishes. Stoves could include electric and hydrogen burners, and ovens would be capable of convec-

A home energy-water system would usually be located in a basement, closet or in place of an unused chimney. Some systems would rely on wind or water turbines, most on photovoltaics. Hydrogen would be generated from greywater plus rain/well water stored in a cistern and/or well with a reversible pump. Many structures would have their own well, but most would share a neighborhood well, often a sewage treatment pond as well. Some would use hydrogen gas for cooking, with electricity a more common option. The clerestory would flood the interior with indirect daylight. Wood stoves would augment this system on the coldest days in remote or mountainous regions.

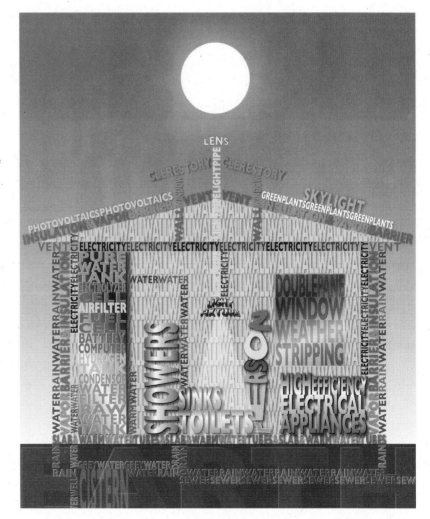

tion or microwave. The homes would require perhaps a tenth of the water and a third of the energy now required, and they wouldn't need furnaces, water heaters and air conditioners.

Drinking water would be 100 percent pure. Since the water is divided into its most basic components and then recombined, all pollutants, chemical and biological, would be eliminated. This process would spawn new businesses. Utilities or recycling companies would buy pollutants collected by electrolyzer membranes, recycling the non-toxic resources while incin-

The proposed energy-water system would make highrises, new or retrofitted, self-reliant on energy they already receive. Downtown areas would maintain a local area grid, allowing buildings that produce a surplus to supply those in shade. Electricity would be generated by photovoltaic wall panels and windows, and by steam generated in rooftop solar thermal units. Water would circulate in columns, capturing the heat of the sun. Buildings would pipe sunlight from the roof to all floors, dramatically reducing energy consumed for daytime lighting and improving the quality of the workplace.

erating toxins. Healthcare firms would develop mineralization cartridges customers could plug in under their sinks, for those who needed water with specific minerals — designer mineralization. Bottled water would seem oddly archaic.

Inevitably some houses would require more power while others would generate less. One might be in the shadow of a massive oak no one wants to cut down, while another might be the residence of a furniture maker with serious power tools. Most homes would include one or two electric cars, and some households might use pick-ups for work. All homes would be linked by a small hydrogen pipeline to equalize output so everyone would have ample electricity and water. The pipeline would effectively carry energy and pure water simultaneously.

Unlike contemporary buildings and transportation systems, which rely on wholly separate energy technologies, a solar-hydrogen-electric system would supplant all existing technologies with one system in a variety of customized formats depending on local conditions. This not only means a home-scaled system could supply vehicle fuel, it also means larger structures could generate sufficient electricity, hence hydrogen, for gas stations and vehicle fleets.

The same hydrogen-electric system used in single-family homes could be used in apartment buildings, office buildings, shopping centers and gas stations. The vast majority of these systems could be powered by photovoltaic roofing and wall panels, but for larger buildings and industries more concentrated energy would be required; wind turbines and solar-thermal collectors could meet the demand.

Solar-thermal power involves various means of collecting and focusing sunlight to generate heat, which can be used to generate steam that drives turbines or Stirling engines that power generators. The technology is low-tech, as exemplified by its reliance on heliostats — sophisticated mirrors — to concentrate sunlight. Today's solar-thermal generating plants are centralized facilities, but smaller units could be manufactured in a variety of sizes.

Big city systems would involve a variety of photovoltaic, solar-thermal and wind technologies, with a local area hydrogen pipeline to equalize output and supply smaller buildings often in the shadows. Office buildings would be sheathed in photovoltaic skins with wall panels and windows generating power. Some buildings would include spires with turbines capturing the power of wind, while others would include solar-thermal generators arrayed around the roof. Older buildings would be retrofitted, often developing rooftop solar-thermal systems to generate heat and electricity simultaneously. Thousands of existing large buildings would just receive new windows and wall panels.

Manhattan would no longer need power plants and distant reservoirs. Cars, subways, elevators, lights and all sorts of other appliances would be powered by photovoltaic, solar-thermal and wind generators so integrated with the architecture of the city they'd be barely noticed, except from the air, where the PV roofing atop of thousands of brownstone row houses would sparkle in the sun. Every home would have ample pure water. Sewage would be disposed of in new ponds set along the rivers and in parks. Existing water tunnels hundreds of feet beneath the city would become reservoirs instead of pipelines.

Freeing the Waters

Every fossil fuel system is producing water vapor not captured by pollution control systems. It wafts from car exhaust pipes, furnace vents and countless smokestacks — visibly on a cold day. There is controversy over the exact role water vapor plays in climate change. Some scientists suspect it may be as significant as carbon dioxide, nitrogen and methane in reflecting heat within the atmosphere.

The new energy-water system could end the continual pumping of warm water vapor into the atmosphere by condensing it back to pure water. This not only means less uncertainty about the atmosphere, but more certainty about the purity of the water one drinks. The new energy-water systems would remove all pollutants, including inorganic chemicals and heavy metals. Billions of energy-water systems could purify the world's freshwater.

The new energy-water systems would eventually supplant centralized power plants, dams, aqueducts and much of a city's water distribution and electrical grid. Homes, office buildings and factories would no longer use water once and throw it away, rather they would recycle most of the water internally, use less for landscaping, and rely on wells replenished by rainwater. In most of the world most of the time this system would provide ample water. Many desert communities have deep aquifers that can be recharged to provide ample storage capacity, but they might want to cut down on irrigating golf courses.

In the US today 48 percent of all fresh water is used in power plants, 34 percent is used to irrigate farmland, 12 percent is used in homes, 5 percent is used by industry, and 1 percent is used by livestock. Water for power plants would no longer be necessary. Irrigation could be replaced by restored groundwater and less water-intensive farming strategies, ending reliance on

Of the water delivered to US households only about one percent, or a little more than a gallon, is consumed as drinking water, or home-brewed beverage. A few more gallons may be consumed in pre-packaged beverages and bottled water. Low-flow toilets, already common, use only about half the water of conventional toilets. Outdoor use varies widely according to region. In the southwest swimming pools are more common, while in colder climates hot tubs may be a major use of water. (Source of quantities: American Water Works Association)

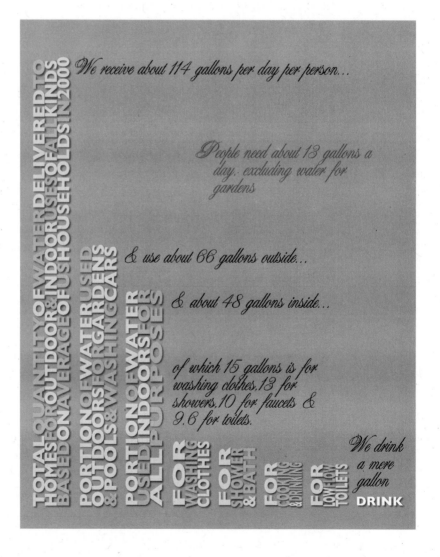

TOTAL QUANTITY OF WATER DELIVERED TO HOMES & OUTDOOR & INDOOR USES OF ALL KINDS BASED ON AVERAGE OF US HOUSEHOLDS IN 2000

PORTION OF WATER USED OUTDOORS FOR GARDENS POOLS & WASHING CARS

PORTION OF WATER USED INDOORS FOR ALL PURPOSES

FOR WASHING CLOTHES

FOR SHOWER & BATH

FOR COOKING & DRINKING

FOR LOW-FLOW TOILETS

We receive about 114 gallons per day per person...

People need about 13 gallons a day, excluding water for gardens

& use about 66 gallons outside...

& about 48 gallons inside...

of which 15 gallons is for washing clothes, 13 for showers, 10 for faucets & 9.6 for toilets.

We drink a mere gallon DRINK

distant reservoirs. Industry can recycle water onsite and reduce demand to where reliance on rainwater and wells would be sufficient. Water no longer used would return to the watershed.

Of the 12 percent used in homes (about 100 gallons per day per person), roughly 48 gallons are used outdoors; 51 gallons are used indoors, in toilets, sinks and washing machines; and one gallon is used for drinking. Gardening would consume far less water if native plants and those well adapted to the local climate were emphasized, along with drip irrigation, mulching and

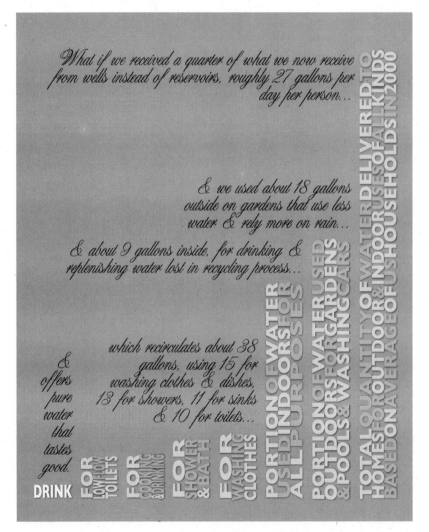

What if we received a quarter of what we now receive from wells instead of reservoirs, roughly 27 gallons per day per person...

& we used about 18 gallons outside on gardens that use less water & rely more on rain...

& about 9 gallons inside, for drinking & replenishing water lost in recycling process...

which recirculates about 38 gallons, using 15 for washing clothes & dishes, 13 for showers, 11 for sinks & 10 for toilets...

& offers pure water that tastes good.

DRINK · FOR LOW FLOW TOILETS · FOR COOKING & DRINKING · FOR SHOWER & BATH · FOR WASHING CLOTHES · PORTION OF WATER USED INDOORS FOR ALL PURPOSES · PORTION OF WATER USED OUTDOORS FOR GARDENS & POOLS & WASHING CARS · TOTAL QUANTITY OF WATER DELIVERED TO HOMES FOR OUTDOOR & INDOOR USES OF ALL KINDS BASED ON AVERAGE OF US HOUSEHOLDS IN 2000

If household appliances now in use were converted to recycling units, or the water was recycled within the building, the quantity of water delivered per household would decline, but we would use the same quantity of water indoors, with nearly all recycled, minus that lost to leakage and by residents going out and taking their pee with them. In most cases and places garden water would drain into the watertable, thus replenishing the reservoir under the buildings.

various shading strategies. In places where gardens were elaborate or water-intensive, wells and cisterns would usually be sufficient, if not ample. Water for washing and toilets would either be recycled in-house or via a neighborhood-scaled biological treatment facility discharging clean water into wells. It would be possible to cut home consumption of "virgin" water by over 85 percent, relying on groundwater as the storage reservoir for the remaining 15 percent, using a combination of individual and community-owned wells monitored by a local water agency.

This strategy would cut water consumption radically, returning most of the water to the land, rivers and sea. Water for power plants would drop to near zero and home and industry use would decline to a fraction of what it is now. Billions upon billions of gallons of clean water would flow downriver, flowing through estuaries and bays, seeping into the earth and carrying the salts of the earth to the sea.

One Generic, A Billion Variations

Each energy-water system would be comprised of one primary power unit, the electrolyzer-fuel-cell with its control systems, and various ancillary components. The primary power unit, analogous to a computer, would be a generic product available from many companies in a wide range of sizes, from 500 watts for small structures to 5 kilowatts for a large house to a megawatt-plus for commercial systems. Just as we now buy computers with peripherals, we'd buy the power unit plus peripherals, including photovoltaics, wind turbines, tanks and software depending on location and building.

Two components would need to be periodically replaced — the electrolyzer and the fuel-cell filters. Each electrolyzer requires filters to remove particulates from water, and as the water is cracked heavy metals and inorganic substances are left behind. They must be reclaimed or destroyed. Similarly, each fuel-cell requires an air filter just as gas engines do, and in the process of filtering air it's possible to extract some measure of carbon dioxide and methane directly from the incoming air. Both gases would be stored in containers, as would water filters, and both would be periodically exchanged at a local store or by a utility company. The gases would be sequestered deep underground.

It may seem more costly to outfit virtually every building on the planet with a new energy-water system over the next quarter-century, but paradoxically, such a strategy would be considerably cheaper in solving *all* the problems we face. Constructing new water and power systems relying on aqueducts and grids, while patching up existing systems, would be enormously costly. The new energy-water system, being inherently decentralized and incorporated in structures, not only uses less of all resources and ends pollution, but also could clean our water and remove greenhouses gases.

This new energy-water system in a box does not yet exist, although a few buildings use such systems as demonstrations. The concept, however, represents a host of emerging trends in energy and water technology. These technologies acknowledge by design the reality that light and rain are naturally given to most of the population, that buildings and cities naturally collect energy and water, and that water naturally flows into the ground just as electricity flows to ground.

PART THREE

Industry
Transformed

8

Light Mobility

THE WORLD'S LARGEST INDUSTRIES have already decided on batteries, fuel-cells, hydrogen and renewable energy. Automotive and oil industry leaders have invested billions in these technologies over many years. Solar-electric transport means nonpolluting electric cars and a gas station in many garages. These new technologies are now being developed for trains, planes, ships, buses, taxis, trucks and cars.

Clouded Vision of Mobility

Transportation in the developed world became synonymous with cars and planes beginning in the fifties and sixties. This dominance has saddled the cities of the world with a landscape where up to two-thirds of the city's area may be devoted to cars, and where heat is not only soaked up by acres of asphalt, but produced by millions of engines idling in the traffic that clogs virtually all urban highway systems. In this way cars add still more heat to the bubbles of warm air that envelop the world's cities and contribute to overall global atmospheric heating.

Paralleling the dominance of cars was the nearly total decline of railways. Between 1950 and 1975 most of the world's railways were either abandoned or struggling on meager capital, with no hope of any assistance. Since then railways have returned to the fore, because they represent the only available avenue to address traffic congestion on the ground and in the sky. Now the railway may also be a primary means of reducing the heat-island effect on vast urban areas, especially those reliant on automotive transportation, while simultaneously reducing use of fossil fuels and overall costs of freight and passenger transportation.

If we're to cope with climate change it is essential not only to reduce pollution, but also to reduce the heat being produced by engines and pavement. However, while these objectives may be easy to grasp and the implications of revitalizing railways may be obvious, there are few major national or transnational initiatives to expand railways dramatically. There are plans in Europe and parts of Asia, notably Japan and China, and even a budding regional plan in the Middle East, but they tend to be secondary to highways, in financing and political priorities, and largely unknown to the public. Yet traffic congestion remains a pervasive and serious problem that is not being addressed on the scale it's occurring.

The US Congress has been debating the status of Amtrak since the federally supported corporation was founded in 1971, and only recently has considered increasing its funding substantially. Evidently Congress hasn't grasped the popularity of railways, both light rail and intercity passenger trains, despite the construction of new railways in over 40 US cities since 1980, and the popularity of Amtrak's Acela service from Washington, DC, to Boston. Worldwide there are many excellent passenger train systems, such as Virgin Rail's high-speed rail operations in Great Britain, the French TGV and other European high-speed rail systems, as well as Japan's Shinkansen or "Bullet" trains. While these systems are certainly a good start, they remain small in proportion to transportation markets, and miniscule in relation to the staggering global impacts of transportation on land, resources and the atmosphere.

Curiously, the potential of railways has been largely ignored by activists, politicians, media, academia and the corporate world. Moreover any public discussion of transportation futures is almost wholly focused on vehicles and new technology, not systems and the realities of transportation. The debate centers around car-energy systems or car pools or car designs or just the car's sexy lines. Economic debates focus on the cost of the vehicle, as if it were unrelated to any larger system. Technological debates focus on speed, as if monorails, flying cars or ultra-efficient airplanes might be magic bullets leading to a perfect world where parking is always free and there is no toll to pay.

The popular transportation story is largely a fantasy. Virtually no one knows what it really costs to move anyone or anything, nor who travels and by what mode. In the US it is widely assumed that everyone lives in the suburbs, and drives and flies; that railroads are dying; and that freight is where the money is. In fact, railways are being expanded all over the world; total revenue generated by all freight transportation is a fraction the size of total expenditures for passenger travel; and trains carry more people between Washington and New York than parallel highways and airlines combined. Manhattan's Penn Station is at capacity and nearby Grand Central just set a record of 75 million passengers a year; about half the US population lives in cities or small towns, about 15 percent doesn't drive and 30 percent *never* flies. The railroads are still here, with over 175,000 miles of track in North America, and they are still the most efficient mode available. There is no other alternative system.

Popular transportation history is largely fiction, invented for the most part by auto manufacturers and their advertising agencies. Evidently their intention was not only to sell cars, but also to eradicate any memory that there was ever any competition.

We are mobile creatures and for most of history our primary transportation investment was sandals. Enclosed shoes was a major technological breakthrough. For millennia people viewed traveling as an occasional activity involving a walk to another home in their village, a walk to a nearby village, or perhaps a lengthy trip by foot, horseback, boat or ship to a far port. Our own two feet were our primary mode of mobility. Few people traveled by river and even fewer traveled by sea. Pre-steamship, a long journey by sea could quickly become an adventure where one's survival was dubious.

For many centuries the tallest structures in any town or city were church steeples. In 1840 most city dwellers in Europe, the Americas and much of Asia lived in one- to four-story structures, in neighborhoods centered around a church or temple. Many more lived on farms, in clusters of homes a short walk from a village with its market and church. Wagon or cart roads linked the cities and towns, but they were often little more than two ruts through wilderness. Commercial stagecoach services were widely available, but cramped, dusty and rough. On the way one might see others walking or riding horses, carts or carriages. Walking a mile or two was normal, a long walk might be 20, even 100 miles.

Everything changed in the 1850s. Railway stations quickly grew to become the tallest structures in many cities, with their clock towers and massive train sheds dwarfing other buildings and all but the largest churches. Instead of a hundred people traveling on any given day there were a thousand,

It is assumed that every American drives or flies. In fact roughly 15 percent of the population doesn't have a driver's license and about 30 percent will not fly — ever. Advocates for public transport are often dismissed with the statement, "Americans won't get out of their cars." No one asks, "…but what would they get into?" There isn't much choice; transit and intercity rail systems are modest at best. Based on the popularity of rail services, an expansion of transit and passenger train services could divert about a fourth of combined airline and automobile traffic, saving energy, money and time, while increasing walking and bike use as well.

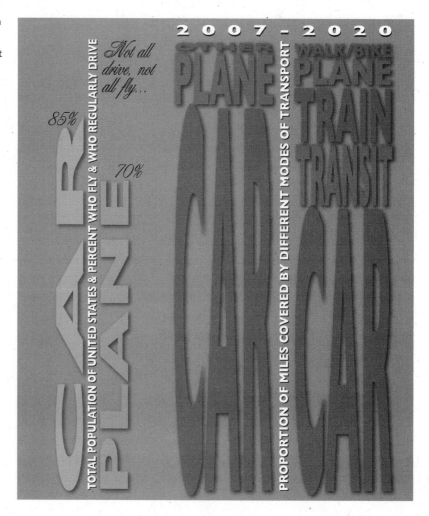

soon to become ten thousand, cruising along at ten times the speed — up to 40 miles per hour. Railways became ubiquitous. In the next twenty years the railway would trigger the biggest change in human genetics ever as millions of people from villages traveled to cities, and met. In the US people whose parents remembered the passing of Thomas Jefferson in 1826 would live to see electric light and telephones, trains going a mile a minute, electric trolleys, subways, pocket cameras and movies. By 1900 virtually everyone rode trains, trolleys and bikes.

Beginning in the 1880s bikes were very popular. Contrary to the automotive advertising fantasy that everyone rode horses in the 19th Century, welcoming the car as a new form of *individual* transportation, horses were costly and not widely used. Bikes set the stage for cars. Cars began appearing around 1900, first electric, then steam and then gasoline, and there were millions of them in the US by 1910. Railways kept right on growing too, building on innovations that made speedy train travel the standard, as exemplified by train names like the 20th Century Limited. Railroads were massive enterprises on the track to success, then came faster cars in the '20s, the Depression in the '30s, subsidized highways in the '40s, and jet planes in the '50s. Railroads were over, or so it seemed.

Between 1940 and 1980 railways lost more than half their track mileage and nearly all their market, while most trolley companies simply vanished. In the same time frame we essentially restructured cities with expressways dividing them like rivers, and one-way streets designed to accelerate traffic to the suburbs while expediting the decline of city neighborhoods. Today the rails carry less than one percent of all passenger business measured by revenue, with airplanes at 12 percent and cars at 86 percent; trucks move nearly 90 percent of all freight. Railways' loss of routes, market share and overall potential has been a global trend. We've all shot ourselves in the foot.

Popular US history of the demise of trolleys is dominated by the National City Lines (NCL) story, ostensibly the true story of how the auto industry secretly plotted to destroy the trolley industry. Unfortunately the story is a tad heavy on the drama and light on the facts.

NCL was a holding company formed by General Motors, plus oil and tire interests. In the 1930s NCL began buying up failing trolley companies to convert them to buses. But NCL's actions were no secret, rather they and their sleek new buses were often welcomed as an alternative to the stodgy trolley companies with their old trolley cars. Then the US Department of Justice took notice and charged NCL with anti-trust violations in attempting to control the bus business. It was found guilty. NCL disposed of the properties, but by then the once profitable trolley companies had tired assets in a shrinking market and were being strangled by subsidized highways and a public convinced anything on rubber tires was better. Cities were the only buyers of failing transit agencies, hence "transit companies" became "transit districts." There wasn't enough business to justify rebuilding the track, so most trolley lines were torn up and a once viable system vanished.

Assuming this pathetic episode was some form of conspiracy to destroy the railways is to deny not just the realities of the specific case, but to skew history to conform to contemporary views. Beginning well before modern auto advertising took hold, in the first decade of the 20th Century, Americans and much of the world became entranced by the car, largely because of the intrinsic value of convenient mobility. The advantages were blatantly obvious. Advertising was little more than an effort to steer the buyers to brands; they were already sold on cars. For most of the last century no one cared about the efficiency of railways anymore than they cared about car gas mileage, safety or even the total cost of driving, which was and remains a total mystery to virtually everyone.

Amazingly, while the US and the rest of the world was mourning the loss of trolleys and trains in the 1960s, the first new railway built in the US or anywhere since the 1920s was under construction in the San Francisco Bay Area. The 70-mile Bay Area Rapid Transit system opened in 1971. That same year what remained of the private passenger train system was taken over by a US-government owned company called Amtrak, and later by its counterpart in Canada — VIA Rail.

In 1980, as if arriving from the Twilight Zone, the first new trolley line in the US since the early 1920s opened in San Diego — the "Tijuana Trolley." Shortly after another line opened in Sacramento, then another in Los Angeles. There are new urban rail lines in Buffalo, Calgary, Edmonton, Vancouver, Baltimore, Cleveland, Pittsburgh, Washington, St. Louis, Portland, San Jose, Atlanta, Houston, Dallas, Miami, Denver, Guadalajara and Mexico City, as well as expanded and rebuilt systems in Toronto, Montreal, Boston,

Philadelphia, New York, Chicago and San Francisco. Curiously the first new railways were built in California, a state where everyone loves their cars.

Transportation is still dominated by automotive transportation. However, the car is no longer the only option worthy of discussion.

Railways are growing for two elementary reasons: cost and space. In one 26-foot-wide corridor a double-track railway can move the equivalent of a 6- to 8-lane highway. In cities already carpeted in pavement covering up to half the land area space matters. The railway can move one person one mile

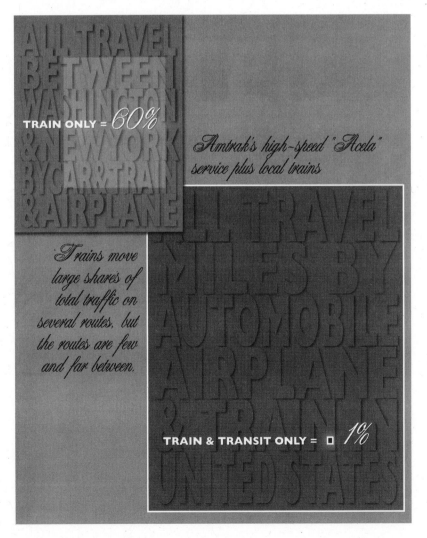

TRAIN ONLY = 60%

Amtrak's high-speed "Acela" service plus local trains

Trains move large shares of total traffic on several routes, but the routes are few and far between.

TRAIN & TRANSIT ONLY = ▫ 1%

Public transportation represents a vital element in any strategy to address climate change, improve a nation's competitive position and reduce dependence on imported energy, as well as respond to high fuel prices and traffic congestion. The single biggest transportation problem, worldwide, is the belief people "love" cars. Based on this belief billions of dollars are spent on automotive infrastructure, while a pittance is spent on other modes. Where there is an alternative, as demonstrated by Amtrak's high-speed service between New York and Washington, trains can attract more passengers then planes and cars combined. People love mobility, not necessarily the vehicle that provides it.

for about 25 cents, while the car costs 60 cents plus. In relation to climate change and peak oil the railway is exceptionally efficient. An average one-ton car might cover about 20 miles on a gallon of fuel, while one ton can be moved 400 miles by rail on one gallon.

In nearly all countries railways are gaining political prominence. Except in the US, where the Bush Administration has strangled federal transit programs and Amtrak's budget while secretly planning two dozen new interstate highways. These predominately north-south highways are designed largely to link Canada, the US and Mexico and accommodate trucks. They are not designed around real transportation problems, rather to facilitate continued growth of some corporations at the expense of all, and the continued expansion of the wallets of a few at the expense of everyone else.

Electricity, Propulsion & Mobility

In a movie called *Who Killed the Electric Car?* the US auto industry, and specifically General Motors, is depicted as opposed to electric cars. This charge is made in response to their "EV-1" electric car, a small battery-electric which was leased to several hundred customers between 1998 and 2003. The car was a brilliant piece of design and lucky customers expressed delight at the car's silence, acceleration and sheer ease of use.

In the minds of many auto industry people the EV-1's battery pack was expensive, its body dangerously light and its range on one charge too short. These and other problems convinced GM the car was not broadly marketable. Its solution, as if bent on simultaneously insulting customers and its employees, was to take the little cars back and crush them like bugs. Customers were justifiably enraged.

Automakers are not opposed to electric cars on principle, but there is an internal industry conflict, with internal combustion and electric factions. There are also legitimate and serious problems, some of which were raised by the EV-1. All automakers are faced with a public that wants them to be innovative, that wants safety features beyond government standards, yet is prone to sue over the smallest error. Companies have become paranoid about product liability, and the EV-1 was very small and very light.

From GM's standpoint, and that of the industry, the unavoidable technical problem was and remains the batteries. They work fine in small cars for limited uses but are not yet viable for a wide range of purposes. The industry wants a means of storing electricity that works for all vehicles, including small cars, big cars, trucks and locomotives.

Dozens of new companies have attempted to build small battery-electric cars, but none have even made a dent in the business. The Tesla, a new car announced in the spring of 2006, represents the most sophisticated attempt. The car and company are named for Nicola Tesla, the inventive genius responsible for alternating current (AC) and innumerable innovations, many shared with Thomas Edison. The car is a two-seat sports car powered by one electric motor using batteries recharged from one's home outlet. It is capable of 0–60 miles per hour in under four seconds — headrests are essential. In any case the first production run was sold in weeks. There are people with the green to buy green.

Within and outside the industry it's widely assumed the debate between internal combustion and electric cars is just an argument over the means of propulsion. It's not. It's about how cars are propelled, produced, designed and even what constitutes a "fuel."

When diesel locomotives were introduced 65 years ago the steam locomotive people thought the competition was about the engine, so they built the biggest and most powerful engines ever. Problem was, the issue wasn't just about the engine, but the entire system of building and maintaining the engine. Diesel locomotives used interchangeable parts, were modular and simple to maintain. They cost more to buy but much less to run. They won. Steam was history in 15 years.

Among electric car aficionados the major debate is over batteries versus hydrogen fuel-cells. New battery technologies, notably lithium-ion and metal hydride batteries, offer shorter recharge times and greater storage capacity, but still take time to charge. Moreover, lithium-ion batteries are known to degrade, thus requiring a costly new battery pack, and they are very flammable. Hydrogen fuel-cells represent a comprehensive alternative because a hydrogen-electric system of the same format would be equally viable for vehicles, even buildings. GM has already developed a home power system using a natural gas fuel-cell; a hydrogen version is feasible.

There are many reasons to develop electric cars. Electric propulsion is more efficient than internal combustion engines. It offers exceptional performance, in acceleration and precise power control, with no pollution and little noise. It simplifies manufacturing, reduces capital costs in machinery and eliminates whole categories of components. Fuel-cells are much simpler to produce than internal combustion engines and have no moving parts. Electric technology also frees designers from the constraints of mechanical components. Driveshafts don't go around corners, but wires do.

GM's Hywire and Sequel concept cars illustrate the potential. The entire propulsion system, including fuel-cell, electric motor, hydrogen tank, com-

A hydrogen-electric car is comprised of electric motors powered by electricity from a fuel-cell, which receives hydrogen from the tank and generates electricity and water. Electricity drives the motors. Water is cooled in a condenser and returned to the water tank. This format allows the car's motor(s) to be reversed in braking, generating electricity to power an electrolyzer, which cracks water to generate more hydrogen, extending the car's range on one tank. Rooftop photovoltaics would provide modest power to keep the car's computer running and provide an electrical boost on hot days.

puter and related heating and cooling systems, fits within an eleven-inch-high platform. The platform looks like a skateboard with wheels at each corner. Any vehicle body can be placed upon the platform: a pick-up, convertible, sedan or station wagon while its controls need only be plugged in.

Several companies have built hydrogen fuel-cell electric vehicles with a total of over 140 cars, vans and transit buses in operation in 2006. These vehicles are primarily in California and overseen by the California Fuel Cell Partnership, an alliance of manufacturers including Daimler-Chrysler, Ford, General Motors, Honda, Hyundai, Nissan, Toyota and Volkswagen. There are 40 hydrogen fueling stations in California, most operated by fleet owners. Meanwhile the US Army, with Sandia Labs, is building a fuel-cell locomotive. Similar programs are underway in Europe and Japan.

Hydrogen fuel-cell systems have problems: cost and service life. As of 2006 the vital membrane where hydrogen and oxygen mix rapidly wears out and is expensive to replace. Plus, the process of producing hydrogen from natural gas, or better yet water, is expensive due in large part to the low efficiencies of certain components. These issues are receiving considerable attention by a small army of scientists, engineers and entrepreneurs.

Hydrogen is a thin gas. Its atoms are far apart. If stored at atmospheric pressure a tank big enough to power a car a few hundred miles would be equivalent to a Mini Cooper towing an Airstream trailer. But if the hydrogen is compressed to 5,000 pounds per square inch the tank's size becomes reasonable. Paradoxically, a high-pressure hydrogen tank may be a safer hydrogen tank. Since it must withstand high internal pressure it's also strong enough to withstand high external impacts.

A high-capacity tank is also possible using the bizarre quality of some metal alloys to soak up hydrogen like a sponge — becoming metal hydrides. This form of tank can hold more hydrogen than pressurized tanks and presents no hazard. Developed by Ovonics Hydrogen Solutions LLC, a subsidiary of Energy Conversion Devices Inc., this storage medium has already been tested in several vehicles.

Another option is glass microspheres. These tiny spheres are used today as a filler in various plastic products. When warmed they absorb hydrogen passing between glass molecules, but when the glass returns to ambient temperature the gas is trapped. Releasing the gas requires heating some portion of the spheres as needed. The microspheres are about the consistency of granulated sugar. There is no risk of fire or explosion.

Curiously, the auto industry regards water exhaust as just a nice feature. Yet at today's prices, spewing pure water is like exhausting money. Moreover, there is little recognition of the potential of water vapor to contribute to climate change; a potential long suspected and apparently confirmed by research on jet contrails.

Hydrogen-electric cars could recycle water on-board via "regenerative"

Existing energy systems are reliant on fuels from the earth, with several oil products used for vehicles, while natural gas, coal, nuclear and hydroelectric power plants are used to generate electricity. Renewable energy, with hydrogen as the prime storage medium, is more practical for all purposes, and thus represents one family of technologies to meet most needs. Notably, photovoltaics are already widely used to power external appliances, such as remote telephone relay stations, billboards and streetlights, which thus need no connection to a power grid.

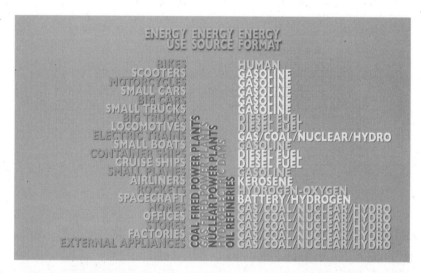

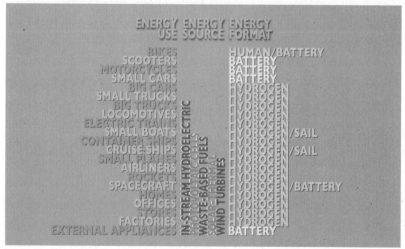

braking. Water vapor produced by the car's fuel-cell when the car is operating at a steady speed or accelerating would be condensed and returned to an on-board storage tank. There would be no exhaust pipe nor water vapor wafting into the atmosphere. When decelerating or descending a hill regenerative braking utilizes the capability of an electric motor to function as a generator and act as a drag on the car while generating electricity to an on-board electrolyzer, a technological cousin of the fuel-cell that uses electricity to crack water into hydrogen and oxygen. Water would become hydrogen again, returned to the storage tank. A car with regenerative braking would have a longer range on one tankful of hydrogen because it would recapture braking energy, otherwise lost as heat off the discs.

No matter how efficient the vehicle's propulsion system it would still deplete the hydrogen. Their roofs could be covered in photovoltaics, extending their range, but vehicles would still need to refill their tanks from time to time. Many car owners might have hydrogen capacity at home, others would go to gas stations as now. Gas pump nozzles would have a very tight seal. The water tank would be emptied and the hydrogen tank filled in seconds.

A few automakers are now introducing cars that burn hydrogen in conventional engines, with microprocessors controlling an ingenious "multifuel" delivery system. Hydrogen can be used as an alternative fuel, and this adds to its allure. Burned in gasoline engines it generates water vapor laced with tiny quantities of burned motor oil. However, unless the water vapor exhaust is condensed back to water and recycled the car would not only add to the warm water vapor already being exhausted by engines, and prone to form clouds, it would also demand a source of water for continued hydrogen production. Water would then be viewed as a source of fuel, rather than a fluid storage medium endlessly recycled.

Reliance on hydrogen-electrics means the water is not a fuel, but a means of storing solar-energy, and as such it's recycled, with modest additions due to leakage. Compared to internal combustion engines the inherent efficiencies of fuel-cells and electric motors mean the quantity of hydrogen consumed and waste heat generated would be significantly less. Vehicle power systems that rely on renewable energy to produce hydrogen from water and recycle the water represent a system relying on the one fuel that drives us all — light. Sunlight, wind and falling water would be the source of energy, hydrogen merely the means of storing it.

Historically more than one military purchase order has started a new industry. Today's military uses photovoltaic cells and fuel-cells. Those who envision tomorrow's military are thinking electric. They might see an eight-wheeled armored vehicle with hydrogen fuel-cell power, variable traction motors, full torque at all speeds, jackrabbit acceleration, on-board pure water and silence. Silence. They envision electric Navy ships, with ray guns that dispense with the artillery shell and replace explosive charges with charged beams. Before you can say, "What the f…," you're toast. These all-too-imaginable horrors will probably be paralleled by a raft of civilian and comparatively benign spinoffs, ranging from self-contained electric recreation vehicles that unfold at the campsite, generating electricity and pure water, to electric construction machines that silently do the job like giant insects.

There will be solar-powered ray guns. There will be solar-powered Formula One cars that break all track records. There will be the guys at the corner gas station who know how to make a stock fuel-cell car perform, and achieve 0–60 in less than three seconds.

Cruising by Light

Ships and planes are going electric. New ship drive systems, and new control systems in airplanes, are electric. New ships and planes are still powered by a diesel or a jet engine, but subsystems are increasingly all electric.

New "Azipod" drive systems, for example, replace a ship's propeller shaft and rudder with a pod containing a large electric motor driving the propeller. The whole pod protrudes down from the ship's stern and can be rotated 360 degrees. New ships use diesel engines driving generators, in turn powering azipod motors. Ships like the Queen Mary 2 can turn inside their own length. Azipod technology can also save 15 percent on fuel.

Ships and boats with hulls and deck structures covered in a skin that generates electricity are a real possibility. In the sun, electricity would power an electrolyzer, cracking recycled water and recharging the ship's hydrogen tanks. Hydrogen would be drawn off as needed to power fuel-cells, in turn powering the ship's electric motors. Most ships could generate a portion of their power at sea. Upon arrival they'd top off with hydrogen generated by stationary hydrogen plants, possibly using electricity generated by harbor tides.

Photovoltaic fabrics will one day be viable for commercial sailing ships and yachts. Existing sailing cruise ships would be propelled when there was wind, while simultaneously generating electricity from sails and the propeller, thus storing hydrogen for windless days. Fresh water too, plus silent motoring.

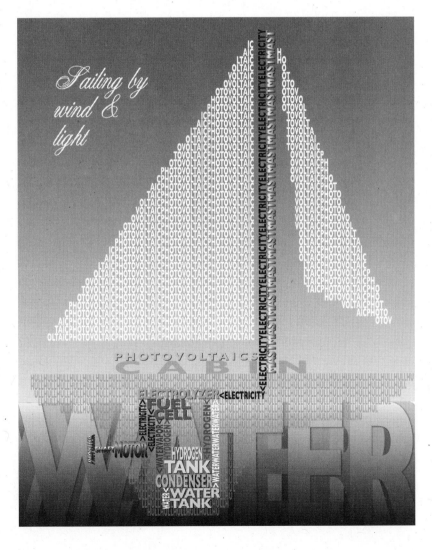

Sailing by wind & light

The energy-water system applied to a sailboat. The boat would never need to stop for fuel or water. Electricity generated by photovoltaics on cabin roof and sails would supply power on sunny days. Water would be recycled, with some portion used for drinking and washing, and losses replenished from sea or lake. When sailing, the electric motor would be driven by the propeller, thus generating electricity. As of 2007 a company called Have Blue in Ventura California markets this concept, sans electric sails. Photovoltaics will likely be integrated with fabric in the near future, thus allowing electric sails.

In 2005 a California-based firm called HaveBlue began marketing a solar-hydrogen power system for yachts. PVs are already common on boats for lights and communications systems. HaveBlue's system uses a photovoltaic array on a boat's cabin roof to generate electricity. The electricity powers an electrolyzer that obtains hydrogen from the water, and the PV output is augmented by the boat's propeller. When under power the electric motor drives the propeller, but when sailing the propeller drives the electric motor to regenerate electricity. The boat can carry 300 miles worth of hydrogen in hydride tanks in the keel. The system is expensive, but the price would decline in mass production. No doubt there are some who'd want a boat capable of going anywhere on Earth with no need to stop for fuel or water, ever.

Planes obviously cannot afford any excess weight, and fuel is already heavy. Moreover large commercial planes consume very large quantities of fuel so rapidly it would seem impossible to power a plane with renewable energy. Yet, paradoxically, planes are deceptively efficient, despite the seeming inefficiency of driving a multi-ton aluminum tube at 500 miles per hour. New executive jets using state-of-the-art engines achieve miles-per-gallon rates that approach automobile efficiencies of around 15 miles per gallon.

Jets could burn hydrogen in jet engines. This would eliminate kerosene jet fuel and its trail of carbon dioxide and other pollutants, but not water vapor.

An electric airplane might at first seem laughable — in the realm of model planes. One group of aviation professionals has already built a battery-electric two-seat private plane. Another innovator is working on a glider with electricity for its small motor powered by photovoltaics on its wings. Model plane hobbyists are increasingly using silent electric motors, not whining gasoline engines.

Boeing and Airbus, the world's two biggest airplane builders, are investigating all-electric planes. The industry is already beginning to replace airplane hydraulic systems for flaps, rudders and other components with electric motors. These planes would still use conventional jet engines. However, electric controls could set the stage for all-electric power.

An electric jetliner would likely use "ducted fans," a propeller in a tube in aeronautical terms. The electric motors would be more compact than a jet engine, but like a jet they'd require significant quantities of energy. A Boeing 747-400 airliner can carry 400 people 8,400 miles on 57,285 gallons of kerosene, or nearly 200 tons of fuel. Using hydrogen to power fuel-cells, which would power the ducted fans, is feasible on board a plane. The question is merely one of batteries versus hydrogen as the means of storing energy.

Batteries tend to be dense and heavy, and many battery technologies are too volatile to risk in such a use. Nevertheless there is considerable research going on in battery technology. It is conceivable a light, efficient and safe battery could be developed for airplanes of all sizes and types.

Hydrogen may be light, but not if compressed in a strong tank. Compared to existing kerosene fuel tanks, high pressure tanks or metal hydrides would almost certainly involve more weight, thus compromising the range of the plane. Moreover, if we seek to reduce water vapor in the atmosphere we would need to contain the water produced by the on-board fuel-cells as they generate electricity.

Perhaps hydrogen could be encapsulated in a water-based gel, a liquid hydride the consistency of syrup. Fuel tanks would be filled much as they are now. As the hydrogen is discharged to the fuel-cell the water produced would take the place of the shrinking gel.

All-electric planes could translate to no risk of explosion or fire, much less noise, and zero pollutants or water vapor. Planes would still leave a trail, a wisp of warm air, but less overall heat due to electric propulsion. Airports wouldn't smell of kerosene anymore, except on special days when antique planes dropped in for a visit.

Return of the Retail Railway

For a century you could step aboard a train in any city on the North American continent and access 23,000 stations just in the US. Enroute you might enjoy excellent food, a conversation with a newly made friend, or the stunning view from the lounge car. The same system also carried every imaginable form of freight, from mail to package express to livestock to gravel to pianos to cars to Sears catalogs and all the products within. Everyone knew they could go anywhere or ship anything via their local station — a retail railway.

There are four transportation infrastructures on the planet that move the vast majority of people and freight — highways, airways, railways and waterways. Worldwide highways move the most, by number of people or volume of freight. Airways are second in moving passengers, but next to last in moving freight. Railways are second in moving freight and third in passengers. Waterways are last in freight and passenger volumes, being inherently limited to rivers, oceans and canals, and relatively slow speeds. However, if these four systems are viewed in relation to one another they now form a global distribution system that's increasingly defined as one seamless supply chain extending around the planet. This whole system represents an

Railways and highways are often compared with no understanding of the fundamental realities of transportation, nor how those realities have been obscured by subsidies granted highways. In fact, railways cost significantly less to build and operate, take up much less land, use a fraction of the energy and offer speed and comfort unattainable by cars or buses. In cities the issue is how many people or tons can be moved in how little space.

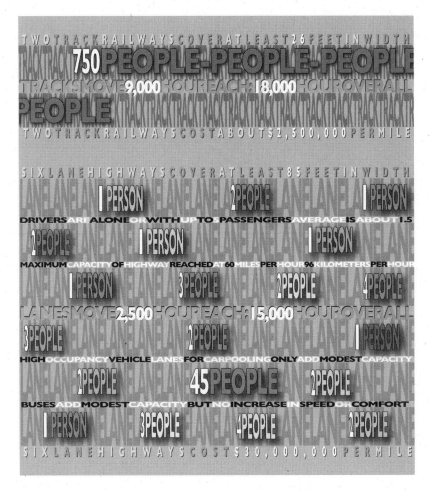

investment in the trillions of dollars, but one now constrained by highway capacity and the physical limits of expanding them. Only one of the four systems can be consistently and economically expanded, and only one offers such dramatic reductions in cost and impacts that its redevelopment would save more than it would cost: the railway.

Railways use less of everything. Compared to cars and trucks they consume a tenth to a quarter the energy, require a quarter the land and utilize equipment that lasts decades. Railways are more efficient by every measure, thus their cost is significantly lower, in many cases 50 percent or more. No one, in over 150 years of trying, has even come close to inventing any mode of transportation as efficient and generally useful.

More than any other innovation the double-stack freight car, invented by a railcar engineer in the early 1980s, typifies the efficiency of the railway. This breathtakingly simple idea involved lowering the floor of a flatcar between the wheels so two containers could be stacked one atop another. This meant the same railyards could handle a train of 200 containers with two employees instead of 100 trailers or containers on flatcars — doubling capacity with no change in track. Double-stack went from idea to reality on hundreds of trains all over North America in ten years. One double-stack train operated by two people replaces 200 trucks with 200 drivers, and the train will keep going in weather that stops trucks cold.

Despite innovations and a measure of growth in traffic since deregulation in 1980, US and Canadian freight railroads still move less than eight percent of total freight revenue. They move 40 percent by tonnage, mostly coal and grains, so they have little impact on lighter freight moved short distances.

Trolley companies and railroads were once corporations aggressively developing passenger and freight business. Now the industry is divided. Freight railroads are run as corporations that pay all their costs, and compete with trucking firms that do not pay the full cost of roads. Intercity passenger service is provided by federal agencies, Amtrak or VIA, and local transit service by city agencies, and they are all strangled in a bureaucratic prison, unable to grow or otherwise function as a business. Transit and intercity passenger railway development remains in slow mode in North America. It's growing everywhere else.

Nevertheless there is a market that wants rail transportation. Proven potential on existing passenger train and trolley routes with dense service can range from 5 to more than 30 percent of total travel volume. Even in car-culture cities rail systems are generally crowded. In California the 100 mile route between San Jose and Sacramento is now at 25 trains a day, and the potential might be several times greater judging from Amtrak's performance on the few routes in the US where it has frequent and fast trains.

Passenger trains and trolleys offer a quality of travel not possible with cars or buses. They can offer certain arrival times, instead of uncertain ex-

Self-propelled railcars are widely used worldwide and are analogous to automobiles because they have a diesel or gas engine under the floor — no locomotive. Here one railcar is compared to a comparable number of automobiles, and expanded to the national potential. Numbers highlighted are generally ignored today, including parking spaces, the value of a driver's time and/or the time spent in transportation. Auto or train driver's time was valued at $50 per hour. Passenger-mile means one person moved one mile (numbers are based on US Department of Transportation statistics).

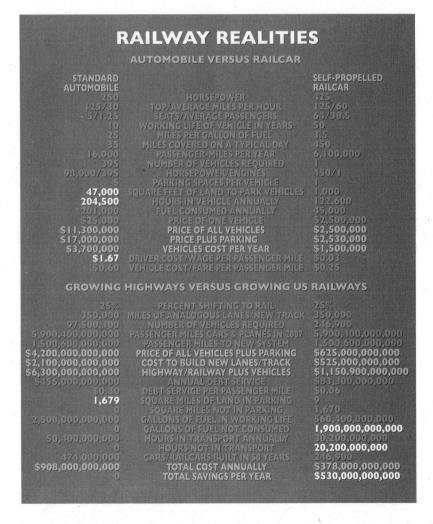

RAILWAY REALITIES

AUTOMOBILE VERSUS RAILCAR

STANDARD AUTOMOBILE		SELF-PROPELLED RAILCAR
250	HORSEPOWER	425
125/30	TOP/AVERAGE MILES PER HOUR	125/60
5/1.25	SEATS/AVERAGE PASSENGERS	64/38.5
10	WORKING LIFE OF VEHICLE IN YEARS	50
25	MILES PER GALLON OF FUEL	3.5
35	MILES COVERED ON A TYPICAL DAY	450
16,000	PASSENGER-MILES PER YEAR	6,100,000
395	NUMBER OF VEHICLES REQUIRED	1
98,000/395	HORSEPOWER/ENGINES	450/1
5	PARKING SPACES PER VEHICLE	1
47,000	SQUARE FEET OF LAND TO PARK VEHICLES	1,000
204,500	HOURS IN VEHICLE ANNUALLY	122,600
201,000	FUEL CONSUMED ANNUALLY	45,000
$25,000	PRICE OF ONE VEHICLE	$2,500,000
$11,300,000	PRICE OF ALL VEHICLES	$2,500,000
$17,000,000	PRICE PLUS PARKING	$2,530,000
$3,700,000	VEHICLES COST PER YEAR	$1,500,000
$1.67	DRIVER COST/WAGE PER PASSENGER MILE	$0.03
$0.60	VEHICLE COST/FARE PER PASSENGER MILE	$0.25

GROWING HIGHWAYS VERSUS GROWING US RAILWAYS

25%	PERCENT SHIFTING TO RAIL	25%
350,000	MILES OF ANALOGOUS LANES/NEW TRACK	350,000
97,500,300	NUMBER OF VEHICLES REQUIRED	246,900
6,900,400,000,000	PASSENGER MILES CARS & PLANES IN 2007	5,900,400,000,000
1,500,600,000,000	PASSENGER MILES TO NEW SYSTEM	1,500,600,000,000
$4,200,000,000,000	PRICE OF ALL VEHICLES PLUS PARKING	$625,000,000,000
$2,100,000,000,000	COST TO BUILD NEW LANES/TRACK	$525,000,000,000
$6,300,000,000,000	HIGHWAY/RAILWAY PLUS VEHICLES	$1,150,900,000,000
$456,000,000,000	ANNUAL DEBT SERVICE	$83,300,000,000
$0.30	DEBT SERVICE PER PASSENGER MILE	$0.06
1,679	SQUARE MILES OF LAND IN PARKING	9
0	SQUARE MILES NOT IN PARKING	1,670
2,500,000,000,000	GALLONS OF FUEL IN WORKING LIFE	560,400,000,000
0	GALLONS OF FUEL NOT CONSUMED	1,900,000,000,000
50,400,000,000	HOURS IN TRANSPORT ANNUALLY	30,200,000,000
0	HOURS NOT IN TRANSPORT	20,200,000,000
474,000,000	CARS/RAILCARS BUILT IN 50 YEARS	246,900
$908,000,000,000	TOTAL COST ANNUALLY	$378,000,000,000
0	TOTAL SAVINGS PER YEAR	$530,000,000,000

cuses; they can often be faster than driving or flying; and they can be safer. One can do things on a train that are difficult to do in a car or bus or plane — such as reading, writing, watching TV and working — or simply impossible, such as going for a walk, visiting a restaurant or bar and even dancing.

Railway revitalization could be as politically popular as a new train around the Xmas tree. Railways represent a means to address traffic congestion, the high cost of transport, the rising price of oil, the need to reduce greenhouse gases and the horrific cost of deaths and injuries each year in car accidents. Railway expansion would also respond to the pathetic reality that no North American city could be evacuated in crisis time, as sadly demonstrated in New Orleans and Houston during the 2005 hurricane season.

Remarkably even those who really do love their cars might support railway revitalization. The car lovers would have more space because "they," all the people who don't want to drive, would be on the train.

What if there were a modern railway network that carried all kinds of people and freight 24 hours a day between every town and city on existing track from Nova Scotia to El Salvador, from Miami to Vancouver? Such a system is defined by the existing continental route structure of standard-gauge track, which parallels all major highways and links all major cities in North America. Today this system does not have sufficient capacity to carry much more freight, never mind passengers, but there is space for new track.

Existing railway routes in North America, including Canada, Mexico and the US, comprise nearly 200,000 miles of route. Several mainlines have been double-tracked in the past decade but most remain single with passing tracks. Most routes are within a right-of-way purchased to accommodate more tracks. A few tens of thousands of former railroad rights-of-way remain intact, ignored or used as trails. In addition there are countless boulevards and highways with wide lanes and median strips that can accommodate two tracks, as new trolley lines demonstrate. A train in a high-occupancy or "HOV" lane would be the highest occupancy vehicle and the fastest, at 100 miles per hour.

What would it cost to expand thousands of miles of existing routes and build several thousand miles of new routes just in the US? If the goal is to attract a quarter of all travelers and increase freight business by say 400 percent and achieve this objective in under 20 years, this would require roughly quadrupling the capacity of the existing 140,000 mile network and adding say 350,000 more miles of new track and routes. At an average cost of $1.5 million dollars per mile it would cost about $525 billion for track, plus another $600 billion for 250,000 railcars, mostly coaches, for a total of $1.1 trillion. This sum, equal to an investment of $75 billion annually for 15 years, could be primarily private capital. The resulting system would generate more than $400 billion in passenger revenue and another $150 billion in freight revenue.

Energy consumed in transport is primarily for passenger transportation and specifically for cars. There are essentially two ways to reduce energy consumed, and the cost and pollution it represents. One, we can expand the use of alternatives, especially railways and local transit. Two, we can utilize the latest technologies to achieve 25 to 50 percent reductions in energy used. Even if the US shifted to nonpolluting energy sources it remains imperative the nation reduce resource consumption to improve its competitive stature.

TODAY-TOMORROW

QUANTITY OF ENERGY CONSUMED FOR TRANSPORTATION IN UNITED STATES

PLANE CAR

TRAIN TRANSIT PLANE CAR

Energy used if public transport expanded

Energy used by most efficient technologies

TRAIN TRANSIT PLANE CAR

A comparable expansion of highways would cost roughly $6 trillion to build and $900 billion to operate. It would require more than 150 times the land, for some 97 million automobiles — 400 for every railcar. Moreover such an investment would merely sustain traffic congestion, not reduce it, while saddling the economy with continued high transport costs.

A railway program would save more than it costs. It would require one sixth the capital and half the operating costs. It could alleviate traffic congestion in many urban areas, save countless hours of time, reduce urban heat

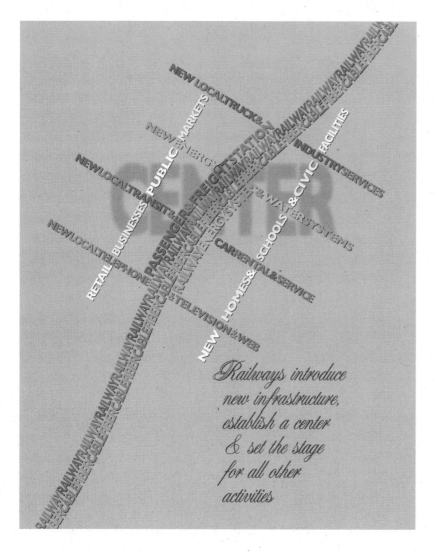

Railways introduce new infrastructure, establish a center & set the stage for all other activities

The construction of transportation systems invariably involves other infrastructure, but railways uniquely require a sophisticated communications system to manage train operations, they often require their own energy system, and if passenger trains are involved there must be water and sewage systems as well. Thus the railway, either all new or a revitalized existing line, is the avenue to introduce new energy, communications and water technologies. It is the one big transport program that provides an opportunity to build all state-of-the-art.

and dust, reduce deaths and injuries and, presuming all solar power, it could cut transportation-produced greenhouse gases by a quarter. Given these implications, one trillion dollars in funding, with say a quarter from public sources, is an incredibly good deal. The money is available. We are already spending it — on driving.

This new system could be a new version of the original — the retail railway. It would be a public transport system that could get anyone or anything anywhere on the continent. It would be a series of regional companies providing all manner of freight and passenger service. The result would be far more freight on railway and far less on highway; far more people getting where they want to go at 90 miles per hour instead of 30; far less money wasted on operation of cars and ownership of second or third cars; and far more fun.

Trains are not just a box to travel in but a civilization in motion. They are of the commons. In this rolling commons, instead of focusing on the highway people are talking, reading, watching the view or a movie, doing e-mail, working, sleeping, writing or having dinner. Trains to countless places would offer a view of a river or canyon or coastline or mountains unseen from the highway. Trains can be a place to be on the way to a place.

Civilization and Transportation Transformation

A century ago there were several railroads built in North America and worldwide, that were essentially built to develop uninhabited regions, after indigenous peoples either died from disease or by design, were starved out or moved out. Invariably the new communities that emerged needed infrastructure. To provide service railroads had no choice but to build water facilities, as well as coal or oil facilities for steam engines, and a telephone system. It was easy to extend these systems into the new community. One US railroad, Southern Pacific, all but developed countless small town economies in the West by building it all and later selling the water and phone systems to cities and utilities. Their remaining company-wide phone system was called Sprint, and was sold in the 1980s.

Building on this precedent the new retail railway could be designed around state-of-the-art technologies. It could be 100 percent solar-powered, incorporating PV cells on railroad ties and roofs, plus solar-thermal facilities over railyards. A network of regional retail railways, with trains powered by light, would not be vulnerable to power outages, oil prices or natural disasters.

Developing new railways inherently involves a virtual encyclopedia of technology. A whole new railway involves track, trains, energy, communications, water and waste systems, plus stations, bridges and tunnels. Also needed are virtually all the professions, from engineers to bankers to lawyers to accountants to architects to scientists, plus of course politicians. Unlike any other large infrastructure the railway involves virtually everything and everybody, a microcosm of the larger society.

A new railway system represents a watershed event. It would not only create innumerable jobs, it would also offer a means for thousands of companies to develop emerging technologies. It would spawn millions of new business opportunities, including at least a trillion dollars invested in real estate projects around stations, with less reliance on cars and less parking. But perhaps most importantly it would act as a seed, with each station being a demonstration of a new energy-water system, as well as new landscaping, lighting and architectural strategies, and battery-electric bikes, and in many cases even new cars, specifically hydrogen fuel-cell electric rental cars. Railways, once the seminal infrastructure that initiated industrialization, may be key to growing the new infrastructure.

9

Making Things by Light

Ecology of Technology

INDUSTRIAL ECOLOGY might seem a contradiction in terms, yet the two words describe a new body of work focused on ecological principles in the design of industries. As a strategy, industrial ecology can be applied to the design of a community or the development of an industrial complex. The concept characterizes the web of exchanges that define how industries and individuals use resources much as one would describe an ecosystem.

One industry's waste is another industry's resource. A century ago railroads commonly removed worn ties and rails from mainlines and reinstalled them on secondary routes. They call it "cascading." This process generated small companies that completed the process, selling worn-out rail to steel mills and worn-out ties to landscapers for garden steps, or to companies that burn them to generate electricity.

Taken as a whole this process is eco-logical not just because it maximizes the use of resources, but because each step in the process maximizes the value of the resource. An ancient railroad tie that once carried the 20th Century Limited is junk to the railroad, a stairway to a gardener.

Fire is the great leveler. There are companies defining a new mode of waste disposal that characterizes industrial ecology. Envision a vertical furnace chamber containing a controlled high temperature mass of waste material starved of oxygen. The furnace takes in all forms of waste — tires, batteries, plastics, glass, cars and toxic materials — and transforms it into molten or gaseous material. Gases are drawn off the top and separated into products, while limestone, steel, aluminum and other products are drawn off the bottom. Around the facility there would be metal foundries, gas and cement companies, all sharing electricity and hot water from the plant. An eco-industrial park would use everything — one way or another.

Technology is generally viewed as a series of discrete products, rather than a system as complex as any ecological system. A major product manufactured for one purpose may be sold, used and recycled, being re-made using the old materials. Waste generated in the process may become a "product" sold to another company, becoming its primary resource to produce yet another product. This activity may trigger additional businesses, such as schools to teach people how to make the product. In industry, as in ecology, one cannot do just one thing — everything is linked.

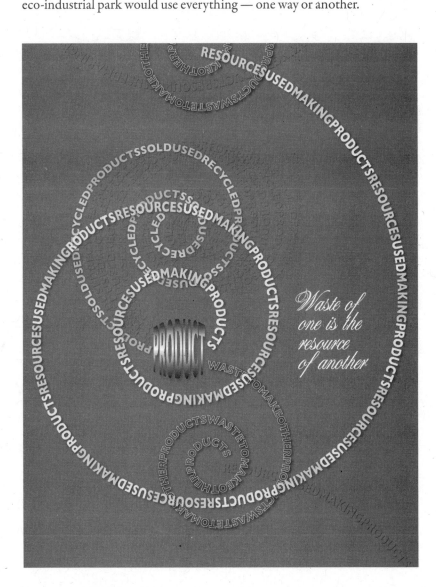

Waste of one is the resource of another

Eco-industrial thinking can also be applied to a whole community of industries. Hydrogen reliance will transform the energy infrastructure. Plastics companies will shift to non-oil sources. This could result in soy-plastic car bodies, as Ford recently demonstrated with the all-soy concept car. The company making the raw plastic might be the centerpiece of a complex of factories, from body-component makers to a junkyard complex where parts are reconditioned or ground up to be recycled.

Industry today is defined by innumerable one-story concrete buildings with flat roofs. These buildings can be converted, or built new, to achieve 100 percent reliance on renewable energy. Some might use a sawtooth roof, as shown here. Some may require more energy, using photovoltaics plus solar-generated steam or a wind turbine. Electricity would be used to generate hydrogen cracked from rainwater, stored in a cistern and/or well with a reversible pump. Pure water, steam and hot water may be used in industrial processes or for heating. This strategy can also include indirect skylighting, and fiber-optic solar task lights, thus reducing lighting costs and improving the working environment.

Industrial ecology is also happening on the micro-level. More and more consumer products are being designed to consider the total environmental impact. The LEED standards (Leadership in Energy and Environmental Design) are a voluntary set of "green" building standards increasingly accepted in the US. These standards place value not just on how little energy a product may use in operation, but on how little it uses from development to the end of its life — its life cycle.

The very idea of industrial ecology naturally leads to modes of sustainability where resources are stewarded as carefully as a gardener would tend his garden. Several retail businesses have sought to achieve a high level of sustainability by minimizing impact, notably the Orchid Hotel in Mumbai, India, where even the toilet paper cores are recycled.

Industry on a Desktop

You can make anything. In the world of modern machine shops it's possible to use Computer Aided Design (CAD), possibly with some Computer-Aided Engineering (CAE) to design a framus. Then you can plan the manufacturing process using a Critical Path Method (CPM) program. Then you

Semiconductors and the photovoltaics were invented in the 1950s but neither became major industries until the 1980s. Personal computers were the first major use of semiconductor technology, and the now ubiquitous PC transformed our world. Computers have since changed virtually all industry. With the emergence of the Web in the 1990s the computer became a gateway to the world's knowledge. Several other trends also emerged, in part as a result of the computer, often growing from the intellectual ferment around high technology.

Cell phones, smart products, desktop production, websites, blogs & you've got mail

SEMICONDUCTOR REVOLUTION: 1985 TO 2000

COMPUTERS SIMPLIFY ALL MEDIA PROCESS
CELL PHONES TRANSFORM TELEPHONY INDUSTRY
COMPUTERS SPEED PRODUCTION WITH QUALITY
MATERIALS MADE ATOM BY ATOM ARE MADE
WEB TRANSFORMS COMMUNICATIONS
COMPUTERS & CONTAINERS GLOBALIZE ALL INDUSTRY
FREIGHT BEGINS RETURN TO RAILS FOR LONG HAUL
ELECTRIC CAR DEVELOPMENT BEGINS
RENEWABLE ENERGY SECURES NICHE MARKETS
FUEL CELL DEVELOPED TO MARKETABLE PRODUCT
ORGANIC FARMING & RANCHING BEGINS
ECOLOGICAL RESTORATION BEGINS WORLDWIDE
LONG HAUL TRUCKING BEGINS DECLINE

1980 1981 1982 1983 1984 1985 1986 1987 1988 1989 1990 1991 1992 1993 1994 1995 1996 1997 1998 1999 2000 2001 2002 2003 2004 2005

can send your electronic files of framus parts to several companies set up for Computer Aided Machining (CAM) and Computer Numerical Control (CNC) systems. Then they will set up their machines, load the metal or plastic or wood raw material and press "Start." Minutes, hours or a day or so later you've got a framus. A "framus" is anything you want it to be in just about any shape imaginable, and you can have just one or dozens or thousands.

Computers have revolutionized the making of mechanical parts. This infrastructure revolution largely happened between 1975 and 1990, and virtually all machine shops, vehicle factories and shipyards worldwide now utilize the technology. Paralleling this shift, computer-aided design programs swept the architectural and engineering fields, as did similar programs in desktop publishing, music and movie production. All these changes have greatly accelerated production, and often improved the quality, of just about everything.

Computers allow us to design an entire "anything" and experience what it would look like before the real thing is built. In the making of the 787 Dreamliner, Boeing Corporation brought designers together with airline

Energy & transport technology set stage for restoration of land & waters

LIGHT REVOLUTION: 2010 TO 2025

INNOVATIVE SMALL ENERGY SYSTEMS FLOURISH
RENEWABLE ENERGY BECOMES PRIMARY SOURCE
URBAN PRIMARY SOURCE BECOMES GROUNDWATER
PURITY BEGINS TO INCREASE WORLDWIDE
PAVEMENT REDUCED BY TWO THIRDS
RAILWAYS TRIPLED IN CAPACITY WORLDWIDE
ELECTRIC CARS & TRUCKS BECOME DOMINANT
ORGANICS TRANSFORM AGRICULTURE
EXPANSION OF PRAIRIE & FOREST RESTORATION
RESTORATION OF MAJOR RIVERS BEGINS
ELECTRIC PLANES SUPPLANT ALL JETS
RATE OF WARMING SEEMS TO SLOW
FOSSIL & NUKE FUEL USE BEGINS DECLINE

2005 2006 2007 2008 2009 2010 2011 2012 2013 2014 2015 2016 2017 2018 2019 2020 2021 2022 2023 2024 2025 2026 2027 2028 2029 2030

The trends that set the stage for the interstate highway and later semiconductor revolutions were evident in preceding decades, just as the trends that could alter our future are visible today. Notably, and contrary to popular belief, ecological restoration can happen very quickly, especially for wetlands and grasslands. Entire fleets of vehicles, such as all the cars in the US, could be replaced with electric versions within 15 to 20 years at current rates of replacement. These trends in energy, transport and land use could result in a decline in the rate of global temperature rise within 20 years — could.

owners, employees and regular customers and they all worked on a 3-D design in color. Such collaborative design is facilitated by computers and the Web, and this virtual infrastructure makes creating the real thing possible in half the time compared to pre-computer methods.

These new technologies make it possible to build cars locally. Manufacturers would design the car and suppliers would ship parts to dealers. Customers would walk in, drive sample cars and then design their cars with a salesperson. It would be assembled the next day. Parts would be modular, with an infinite number of possible design and color combinations. This would trigger major growth in the custom car business.

It is possible to buy a 22-foot shipping container, add some windows, skylights and a porch, plus some photovoltaics and a satellite dish, a battery pack for power storage, a few computers, printer, cell phone and minor components. One could put it anywhere. A business in a box.

Utopian Infrastructure Revolutions

Business is a collection of agreements as subtle in language as they are complex in content. In between all the trivial talk and trillions of numbers buzzing over the Web there are a billion ideas growing. One e-mail leads to four, which leads to an outline, which leads to a proposal, which leads to investment, which leads to market, which leads to a problem being addressed in a fresh new way, which leads to thousands then millions of employees and customers. Just a bunch of agreements. They can be made with lightning speed.

There is an unprecedented coalescence of information happening on the Web. The majority of the world's major libraries, plus millions of corporations and government agencies, are increasingly represented on the World Wide Web and linked to one another and/or countless citizens via the Web. The Web, as a repository of knowledge, is equivalent to a global library of incomprehensible scale.

Given this resource, as well as global networks of service groups, industry groups, business associations, corporations and universities, there is no doubt all societies together possess sufficient knowledge and capability to address the common global problems we face. Given the extent of information on the Web, and the prospect of collaborative design and computer-aided manufacturing, it becomes quite conceivable that new infrastructure could be built very rapidly over a large area with uncommon quality.

Business is a bunch of agreements. The Web means billions of people can and will coalesce into ever larger and more powerful groups and they will continue to build what is now just a loose collection of initiatives scattered widely over the planet. They will do so because this infrastructure revolution represents a response to the first threats truly common to all humanity — peak oil and climate change. This infrastructure revolution is already growing from a billion points at once and guided by a popular vision of sustainability via technologies that weigh lightly on the land and our lives. This new infrastructure demands our businesses reach for utopia.

PART FOUR

Life Revitalized

10

Restoration, Renewal and Rebalancing

The Imperative of Ecological Restoration

TODAY OUR CHOICE as a species is to continue using fossil fuels and accept both the economic and ecological uncertainties of doing so, which may include dramatic climatic change, or to seek methods of ending reliance on fossil fuels that also result in a reduction of greenhouse gases already present in the atmosphere. Moreover, it would seem vital that we accomplish these changes while simultaneously addressing a raft of problems surrounding the use of land, water and energy. There are three key strategies: first, replace the use of fossil fuels, as well as the output of warm water vapor; second, increase the area of green plants; and third, if necessary, use mechanical means to extract and sequester specific gases.

These actions, to some unknowable measure, would result in the rebalancing of the atmosphere and a return to a slightly cooler world. We cannot know the precise rate of climate change now. Therefore we cannot know the precise effect our actions to ameliorate the situation may have. Clearly we cannot replicate what was, but we may be able to diminish our impacts so rapidly we slow and then reverse the trend. We can be sure of one thing, plant life is vital.

Nature is a powerful partner. Studies by a variety of groups reveal the inherent efficiency of the wild. Native prairie grasses in Kansas are so dense with grass species they take in and sequester more carbon dioxide and nitrogen than any farm crop, and can grow faster than any forest. Wild fisheries are far more productive than any fish farm, and estuaries, marshes and wetlands are both vital sources for nutrients and prodigious consumers of greenhouse gases. Grasslands can often produce more protein per acre if they are supporting native species such as buffalo and/or antelopes instead of cows, because these species co-evolved with the native grass and climate. Could we restore grasslands, forests and rivers to radically increase the area and density of living plants consuming and sequestering greenhouse gases,

In most cities the water's edge in the 1960s was a place of old industrial waste and garbage, and a new interstate highway spewing pollution. The railway was in poor condition, the station a pigeon waiting room and the entire waterfront all but forgotten. Downtown was often a ghost town of old offices, empty stores and parking lots.

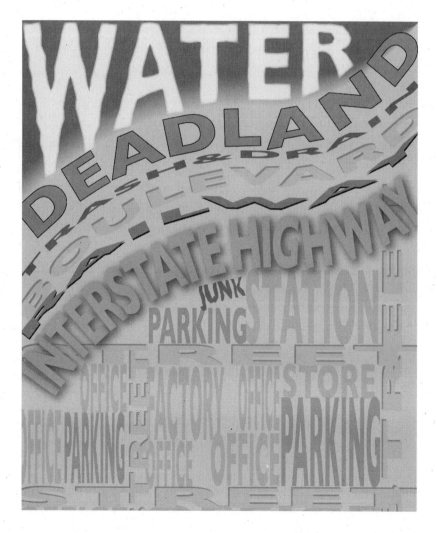

while simultaneously restoring open range animal husbandry, including a mix of wild and domestic ruminants, as well as wild fisheries of all forms, from river trout and catfish to bay clams and ocean tuna?

The central message of ecology is that we cannot do just one thing. The central reality of technology is that each machine does do one thing. The unavoidable reality of our situation is the imperative to accomplish several things simultaneously, to define projects and technologies that accomplish many tasks simultaneously.

Nevertheless, many people are trying as hard as they can to do just one

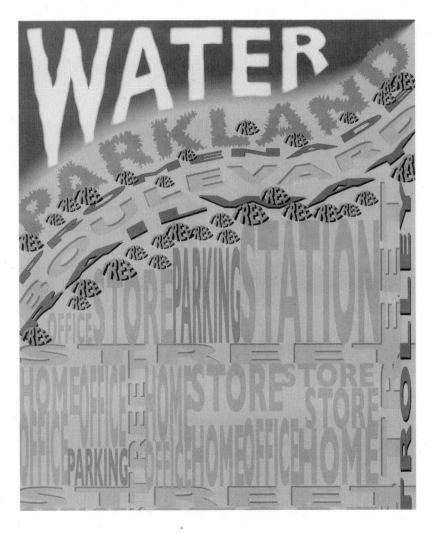

In many cities the water's edge in the 1990s was a place of restored beaches and trees, where the highway is buried or gone. The railway is in excellent condition, the station is busy with people and the entire waterfront is a park. Downtown is filled with restored and new buildings, stores, homes, offices, fewer parking lots and a new trolley line.

thing, with potentially horrific implications. In the urgency to produce one thing, ethanol, to address one problem, the need of an alternative fuel, governments encourage the clearing of savannah and rainforests. Places where hundreds if not thousands of species live in complex and very productive ecosystems, places offering many species of fish and fiber, grain, root and fruit, as well as unknown treasures of DNA in species as yet unnamed. All this is destroyed to produce sugar cane, which is largely used to distill ethanol for vehicles. To do this one thing billions of things are undone. Billions of ecological relationships, a precious legacy of information that evolved over millions of years and a complex of living organisms that consumed carbon dioxide are sent skyward as more carbon dioxide, only to be replaced by one plant to do one thing.

Ethanol is a major source of fuel in Brazil, and is growing popular in the midwestern US as millions of acres of corn are devoted to ethanol production. The final product is then mixed with gasoline. Ethanol and other bio-fuels are a disaster in the making precisely because they represent a single-minded pursuit of a single objective with little or no consideration of anything else. Whether the objective is to produce ethanol from crops such as corn or sugar cane, or to grow seed oils, such as hemp or other oil rich seeds, the notion that the world can grow sufficient fuel without exacerbating the problems of agriculture, notably water diversion and desertification, is absurd. Expanding farmland to produce fuel would almost certainly result in further destruction of grasslands and forests, both critical in withdrawing and sequestering carbon dioxide. Expansion of biofuels also places fuel in competition with food, and that would very likely mean rich nations would commandeer prime cropland, while poor nations end up with little or no food or fuel, and we all lose more rain forests.

Oils that naturally grow in the seeds of several plants are very valuable. It would seem sensible to expand the use of such oils to replace petroleum as now used in plastics and various industrial purposes, but not in such quantities as to replace gasoline, kerosene and diesel fuel. Bio-*fuels* represent a unsustainable strategy that will merely reduce greenhouse gas production, while complicating several other issues related to land use and economy, while bio-*oils* represent crops and quantities that can fit in with the production of other crops, or be grown in a wild state on a sustainable basis.

All of a Piece

New Orleans was severely damaged by Hurricane Katrina and virtually all agencies and citizens agree that if such losses are to be diminished in the future, downstream wetlands must be restored. Prior to the hurricane there already were local groups engaged in studies and test restoration programs. All along the Mississippi River there are groups involved in river restoration; in the summer of 2006 a Google search for "Mississippi River Restoration" returned 2.8 million hits. A US Army Corp of Engineers site was summarized: "Chief of Engineers Recommends Ecosystem Restoration and Navigation Improvements for Upper Mississippi River and Illinois Waterway."

Restoring these wetlands translates to actions that will increase the flow of silt south of New Orleans enough to cause the wet *lands* to rise and expand out to sea, rapidly followed by the growth of stabilizing plants that thrive where salt and freshwater mix. This dense matrix of trees and salt-tolerant plants would become a buffer between the city and the sea.

The wetlands of the lower Mississippi are a gift of the entire river and all its tributaries between the Rocky Mountains and the Appalachians. If we go upriver from New Orleans we'll find the needed silt, millions of tons of it behind dams like deposits in a bank. If we go down river beyond the swamps and go deep into the Gulf of Mexico we come to a dead zone wrought by two centuries of dumping human, farm and petro waste. This dead zone encompasses a swath of undersea territory on the outward slope of the Mississippi River's alluvial fan, precisely the landscape that once received a veritable rain of organic nutrients carried by the river from all points north. The community of sea creatures that once flourished in this rich environment was effectively poisoned. There, right where the river deposits a cornucopia of organic nutrients from all points north, the ecosystem is dead.

Restoration of the Mississippi River system would be expensive, but spread over such a vast area the cost could be manageable and the benefits immeasurable. Many dams and locks would be removed. Barge traffic would decline, replaced by expanded railways. Thousands of smaller levees would be removed and hundreds of floodgates installed. Large areas of cities would be surrounded by levees, and other areas would often be transformed into two- or three-story neighborhoods with open first floors to accommodate the occasional flood. Highways, railways and/or trails would be atop virtually all large levees, serving small communities and farms rebuilt on mounds. Houseboats would be more common, often forming into floating communities migrating to follow the fish or the work. Millions of square miles of farmland would receive floods and vast wetlands would explode in green. Gradually the populations of every species from shrimp to 'gators to crawdads to catfish would rise.

We can view the entire river system not just as a source of silt and nutrients, but as a fishery extending from northland woods to the Gulf of Mex-

River environments are being restored in several regions, often where they pass through cities, as in the Twin Cities and Houston, and where riverside environments have retail or recreation value to local residents. In addition to contributing to groundwater and fisheries restoration a new river can act to cool the community, provide a recreation amenity and create a riverside view for residents living nearby. The river is no longer just a drain, but a living system.

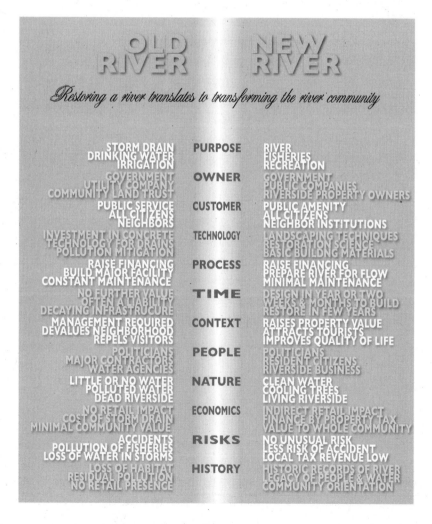

ico; a source of fiber and wood; a pathway of commerce and culture and a means of re-establishing the bulwark of Gulf Coast communities, the wetlands, swamps and coastal grasslands. Even if we were unsuccessful in staving off a rise in sea levels of a foot or more, and the inundation of most of the coastal communities, river restoration would still facilitate the growth of vital fisheries.

New Orleans 20 or 40 years from now could be a vital city at the center of a vast restored estuarine community. Beyond the levees thousands of small delta and wetland communities would be on mounds or stilts. The

Floods are a gift to land & sea... New Orleans is a gift of the Mississippi

Controlling flooding with dams, despite fish ladders and hatchery programs, often resulted in stagnant reservoirs, blocked silt, far less life, coastline erosion and saltwater intrusion. This action, combined with overfishing and the loss of marshes, led to a decline in native species from river to sea. Coastal estuaries are nurseries for many fish, so a loss of marshland diminishes ocean fisheries. If coastal regions are to cope with rising sea levels and further saltwater intrusion, and if fisheries are to be restored, river restoration is imperative.

ocean would be higher, the coast often further inland with swamps expanding to the north, yet a veritable flood of new silt, long stored upriver, would have expanded coastal estuaries around New Orleans, with some used to raise portions of the city.

A major city has already raised itself. Chicago raised itself four to fourteen feet in order to rise above the water table which had made the streets nearly impassable sinks of mud or ice. Major buildings, some a few stories and made of stone, were jacked up and given a higher foundation then all the ground around the building was raised with fill until the sidewalks and streets were as high as the buildings.

It took 20 years, starting in 1855, and during that time most businesses just kept doing business even while crews of laborers were turning jacks to raise the store.

The Mississippi could be a lively place, but the river traffic would be more about fishing and tourism than shipping. Oil — from new sources — would still be big business, serving new plants just north of New Orleans. Soy, cottonseed and hemp oils would be made into biodegradable resins for a variety of products. Revenue from the new oils would finance the cleanup of petrochemical plants. Upstream for several hundred miles countless acres of farmland would be flood plain again, often with creeks and sloughs redefining the farmland into pastures of dense grasses and shrubs alternating with bands of riparian woods, causing a dramatic growth of plains elk, deer and antelope as well as several species of fish all along the river. Sternwheeled riverboats would still whistle for the bridges.

Farm as Garden

Oil and all its byproducts are essential in agribusiness, but optional in agriculture. Machinery can be powered by renewable energy. Fertilizers and pesticides can be replaced with organic substitutes and by farming strategies that don't rely on external additives. Such a change is neither prohibitively costly nor technically difficult. Amish farmers in Pennsylvania have been working with little or no oil since before there was oil, and according to the US Department of Agriculture they are among the most efficient farmers in the world.

Today's agribusiness, with its monoculture strategies emphasizing vast fields of one crop plowed by farmers riding massive tractors with stereo systems represents not just an unsustainable dependence on oil, specifically about 17 percent of US oil consumption, but also a profoundly primitive and wasteful mode of farming. Farming is stuck in engineering, where farms are factories and fields are assembly lines, instead of biology, where a farm is an ecosystem and fields are living environments.

Agribusiness rests on hubris. It's structured around the assumption that we can control nature. To assume we can is to follow a path of endless uncertainty. Monoculture farms rely on imported fertilizers, pesticides and hybrid crops designed to fit specific conditions. The result is grains, fruits and vegetables that are increasingly divorced from their organic counterparts, as if we knew better than thousands of years of evolution represented by the original plants and the microbes in the soil.

Agribusiness must be transformed beyond oil. Current trends in biology, computers and robotics are bound to intersect and spawn whole new modes of farming. A kind of high-tech, low-tech and no-tech agricultural revolution where tools are designed to fit plants, instead of plants being redesigned to fit tools, and where customers choose the economically sustainable instead of the ecologically unsustainable.

The new farm would resemble a garden. Instead of straight furrows regularly plowed, the fields would be lined with wide garden beds following the contours of the land. Each semi-permanent bed would harbor several species of vegetables, fruits, grains and flowers, all planted in patterns to maximize symbiotic relationships between species. This planting strategy would produce more, yet require less water because the soil would not be exposed to sun. Farms would achieve greater yields with a greater diversity of crops.

Organic farming is often dismissed as labor-intensive. But technology already in existence suggests smaller and lighter machines could take the labor out of intensive. A small four-wheeled vehicle, for instance, could straddle a garden bed while its various robotic arms and sensors plant or pick, cut or pull all manner of crop, with or without a human operator. It could pick that perfect beefsteak tomato behind the squash leaves and gently deposit it in a packing box. And it would be powered by the same fuel that grows the food — the sun.

Organic farming is built around the premise that soil is a living ecosystem. Organic farms, now a rapidly growing segment of the agricultural economy, represent both a return to abiding nature and a future where farms mimic nature. Indeed, there are groups imagining new concepts in farming designed to emulate wild ecosystems — wild agriculture. There is a movement in the US to create organic gardens as home landscaping. Several commercial buildings have featured vegetables and fruits as elements of their landscaping. It's called "edible landscaping."

Contemporary agriculture is based on mining the soil of nutrients, poisoning it with pesticide, flooding it with inorganic fertilizer and often irrigating it in ways that waste water. Organic permaculture and bio-dynamic strategies offer an agriculture that can yield up to ten times the produce using no imported fertilizers, no poisons and a fraction of the water. These results are achieved by creating semi-permanent beds of mixed crops, where the garden bed represents a mini-forest of plants that benefit by symbiotic relationships. This strategy also results in soil rarely being exposed to the sun, thus preserving organisms and saving water.

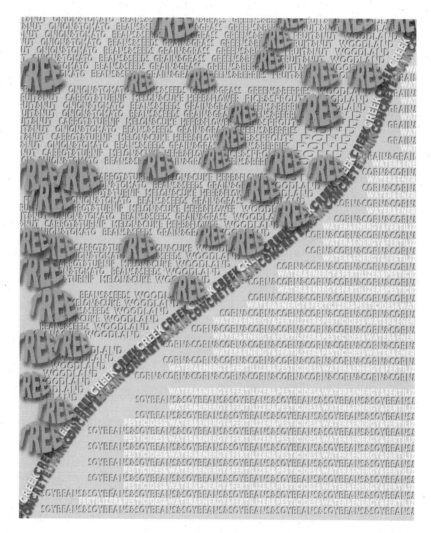

Permaculture, a concept developed over the past few decades and originating in Australia, is a farming and land management strategy where the objective is to create a sustaining landscape where crops for food and fiber, trees for wood and fruit and habitat, and flowers for many reasons and seasons, are woven into a complex ecosystem. It is also a landscape design strategy, where land and watercourse may be graded to maximize the capture of silt or the formation of creeks and woodlands. Permaculture is based on a recognition of nature's tendency to grow complex biotic communities and how to go with that flow.

A region of farms all adopting permaculture strategies could be a garden attracting visitors. The region would sequester considerably more carbon dioxide and nitrogen than conventional farms due to the density of green in fields. It would reduce the heat otherwise reflected off plowed fields.

Celebration of Floods

In many regions, notably parts of California's Central Valley, portions of major agricultural areas in the southwestern US, as well as similar semi-arid lands in Asia, Africa and South America, conventional farms have all but used up the organic nutrients accumulated in the soil over millennia, and sucked up much of the groundwater. Organic material is supplanted with inorganic fertilizers made with oil and natural gas, while the water table isn't replenished because flooding is blocked by dams. These conditions have often led to desertification and land subsidence.

Paradoxically many of the dams and levees built to prevent flooding have all but guaranteed flooding. Before the dams, floodwaters spread out over vast wetlands, causing a springtime explosion of grasses as the waters receded, and leaving silt and recharged groundwater. Dams stopped the floods. Farms often continued drawing water from wells and this often resulted in land subsiding like a giant sponge. Simultaneously the silt that once spread over the land settled in the river channel, raising the riverbed and the river.

Portions of California's Central Valley, and similar areas along the southern Mississippi Valley, Gulf Coast and Atlantic Coast, and hundreds of analogous coastal valleys worldwide, are now more vulnerable to flooding from the ocean than from rivers. When driving levee highways in winter, river water may be lapping the pavement on one side with a 30 foot drop to dry farmland on the other side. Wetlands once several feet above sea level are now at or below sea level. The combination of higher sea levels due to global warming and higher rivers due to siltation, sets the stage for massive flooding of lowlands where the water would be too deep to just pump out. Roughly 1,000 square miles of California's Central Valley is vulnerable now — all around the cities of Sacramento and Stockton. One major earthquake could cause a flood of epic proportions.

Paradoxically in attempting to prevent floods we may have often increased the risk of floods. Levees built to confine the river, combined with reservoirs, resulted in less water flowing downriver at a lower velocity, hence more silt remained in the channel, thus raising the riverbed. Meanwhile farming in deltas and wetlands often consumed the topsoil, causing the land to subside. In lowland floodplains this situation represents a significant risk as sea levels rise — a risk often compounded by upstream streets and roofs causing more storm runoff.

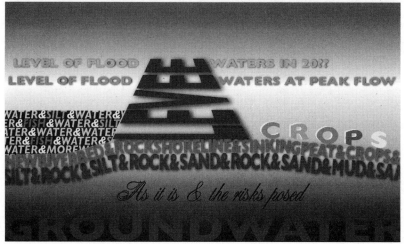

Meanwhile, in more arid farmlands the prevention of flooding is causing desertification, defined as increased alkalinity of soil due to a loss of organic material. Floods caused accumulated natural salts to percolate into the deep water table and migrate to the sea. If not flushed the salt burns the roots of any crop. This same phenomenon is happening in many coastal farming areas, only for a different reason. Withdrawal of groundwater has allowed saltwater to seep in from the sea. In either case farms become a landscape of tumbleweed and sage where lines of white crust encircles stagnant water.

We need a strategy that addresses all the problems of farming, from reliance on imported energy to soil loss to desertification, as well as climate change. Today some communities have reversed land subsidence by injecting water into wells; some farms use no-till strategies; and some government agencies have developed programs to check desertification. Each of these actions is a valuable tactic, but there's no strategy.

Rice growers in California's Sacramento Valley developed a multifaceted strategy that exemplifies how to address many problems at once. For decades they burned the stubble left after the rice harvest, creating clouds of acrid smoke. Forced by air pollution laws to seek alternatives, the growers decided to flood the former flood plains and welcome ducks and geese cruising by on the Pacific Flyway, a sort of Interstate Five for migratory birds. The birds had lost many of their wetland rest stops as people filled wetlands with development. Now they have a new rest stop, the growers get their stubble devoured and fields fertilized, and local hunters get duck for dinner.

Floods can be a windfall not a disaster. Paradoxically, a return to the abiding patterns of river flooding may often be the best strategy to prevent what could otherwise be permanent flooding from the ocean. Floods can be an effective means of rejuvenating the land and restoring the water table. Strategies for flood plain farming could be designed to accept seasonal flooding in order to capture silt, flush out the salts, improve the soil's permeability, restore the water table and increase the volume of plants absorbing greenhouse gases. As the water table is restored farms could increasingly rely upon rain and well water, ending reliance on distant reservoirs. In coastal areas intentional flooding could reduce the extent of ocean flooding due to rising seas by causing the land to rise with the water table.

Renewable energy, using hydrogen as a medium of storage and water as the source, can supplant existing energy sources and cause a dramatic decline in water consumption for power plants. These changes, plus groundwater restoration and low-water farming strategies, would nearly eliminate the need for reservoirs for city or farm irrigation. Dams would be obsolete.

Dams are popularly seen as sacrosanct. Justified by the demand for flood protection, hydroelectric power, urban water and often irrigation, dams are seen as permanent. But dams are just tools. Many dams may be inefficient in proportion to their apparent size: Arizona's Glen Canyon Dam provides only about two percent of electrical demand in the southwestern US, and Lake Powell may lose more water to evaporation than it saves.

Several US dams, from New England to California, are now being considered for removal by local agencies and citizens' groups. A few dams have already been demolished and many more are likely to be removed because the reservoirs are so filled with silt they no longer hold sufficient water to meet irrigation or hydropower demands. Many more dams may be removed because the river is worth more alive than dead. Fish could be worth a whole lot more than electricity. There are historical accounts of creeks, from the watershed of San Francisco Bay to the tributaries of the Columbia, Yukon and dozens of other rivers, where the spawning salmon were so dense one could almost walk across the water on their backs.

Unprecedented Eco-Restoration

Viewing an ecosystem as we would a business, by analyzing the inputs and outputs of all resources, reveals nature is very efficient. Wildlife biologists in Africa in the sixties analyzed the productivity of wild animals versus cattle and discovered a mix of wild animals represented more protein per acre. Best of all, native animals evolved with the plants, so they can subsist on native plants with no help from us.

Many ranchers have discovered the wild. Bison herds once roamed the plains in herds so large they might take half a day to pass by. The many local herds totaled perhaps 50 million or more from Manitoba to Texas and from the Rocky Mountains to the Appalachians. They often roamed in tall grass nearly over their heads, along with elk, deer and pronghorn antelope. Between 1825 and 1880 they were slaughtered by cavalry, settlers and hunters, leaving a mere 500 animals by 1900. That population has since grown to 250,000 plus, and most are owned by ranchers in the high plains, including Native Americans. There's even a bank that loans money to buy bison. The local banker is "Buffalo Bob."

Wild ecosystems are food and fiber factories without walls, managers or machines. If we viewed the entire environment as a productive ecosystem, and sought methods to encourage its productive growth, we could achieve a sustainable harvest of far greater value — in terms of money and creativity — as a fisherman or hunter, than as an employee of a fish farm or feedlot. A wild ecosystem can be a source of considerable wealth in its economic and

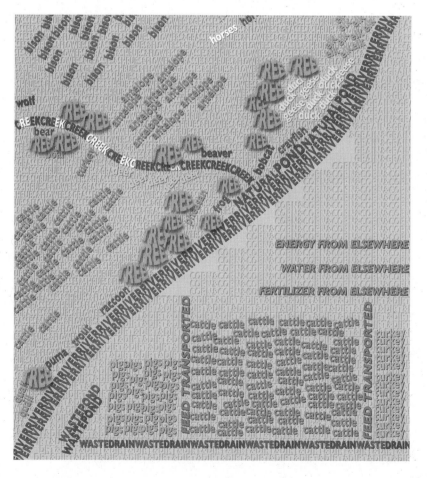

It is popularly believed domestic agriculture is more efficient than wild ecosystems. We grow feed in one place, produce fertilizer in another and get the water from another, all to concentrate animals in production facilities. By contrast in a wild ecosystem the animals, food, water and fertilizer are in one place, functioning with no maintenance and little management, and potentially producing more protein on with fewer resources. Wild grasslands also sequester considerable quantities of greenhouse gases, and grow much faster than forests.

cultural values. This is especially relevant with fisheries, and one fish characterizes this reality more than any other — the salmon.

Salmon spawn in tiny creeks, journey to the sea and return to the place of their birth years later. But due to dams interrupting their path, reservoirs flooding spawning grounds and decades of overfishing, the wild salmon catch is a fraction of what it was. Salmon are symptomatic of the status of fisheries, where a way of life based on the wild is being supplanted by an industry based on machines, managers, fish farms and fish food.

What have we gained by turning a vast wild system involving dozens of species into a fish farm raising one species? Wouldn't it make more sense to restore wild fisheries? Wouldn't our lives and our economies be richer by far if we restored the bays, estuaries and rivers that border many US cities into ecosystems harboring shellfish, such as clams, oysters, crabs, shrimp and crawfish; a wide assortment of ocean and freshwater fish, from perch to bass to salmon to sturgeon; as well as ducks and geese?

Ecological restoration is happening even in urban waters. Wetlands, marshes and shorelines around New York's East River, from the Brooklyn waterfront to Long Island Sound, are being restored, as has been much of the Hudson River. Salmon have been "taught" to spawn again in the Connecticut River. A riparian woodland is coming back to life between the Twin Cities of Minneapolis and St. Paul. Salt marshes are to be restored aside oil refineries in Los Angeles and within sight of Silicon Valley along San Francisco Bay. Some Niagara Falls residents wonder what it would be like if the half of the river's flow diverted to hydroelectric generators stayed in the river, thus *doubling* the volume of water over the falls.

Ecological restoration can include new industries based on the sustainable yield of wild ecosystems. In this new economy industrial and ecological thinking merge. For example, we all need less carbon dioxide and the petrochemical industry needs an alternative for oil. Marginal lands could support hemp and other hearty plants that produce oil, consume large quantities of carbon dioxide and help rebuild the soil. One day there could be photovoltaic factories making plastic roofing from resins derived from hemp oil extracted from plants grown in a floodplain. This example of ecological restoration would be possible because photovoltaics supplanted the dam, and necessary because we need to reduce greenhouse gases.

Ecological restoration is critical to improving our health and increasing the area of green vegetation. We need the protein from wild sources because they carry the complexes of nutrients and oils and minerals we've known for millennia. Wild protein is also more efficient to produce. Tastes good too.

Return of the Condor

The Los Angeles River was once a real river, albeit a creek in the dry summers. As the city grew through the mid-20th Century the area of pavement and roofs resulted in winter rains becoming mini-floods. In response the city transformed once lively watercourses into concrete drains. The LA River is the biggest drain, so big it's been the backdrop for innumerable movie car chases. After a heavy rain the dry concrete channel becomes a brown river a few blocks wide and 40 feet deep. The water races directly into the Pacific Ocean — down the drain.

Restoration could start in the mountains that surround LA, up where the Hollywood sign stands, and down in the canyons where generations of movie stars have lived. It would start with details like driveways, parking lanes and parking lots converted to permeable paving, such as concrete panels with small holes where grass grows and water is captured. The region's parking lots could represent a resource if asphalt is replaced with permeable paving to create thousands of acres of grassy fields. As the area of permeable pavement rises the volume of runoff would decline, thus allowing flood control channels to be returned to real creeks. As groundwater levels rose many creeks would be flowing year-round and residents would notice an increase in wildlife.

Cisterns are decentralized reservoirs. They remain widely used in the world's arid regions. A century ago San Francisco installed cisterns under intersections to provide water for firefighting in the event water mains were severed in an earthquake, as they were in the earthquake and fire of 1906. Combined with wells capable of injecting or withdrawing water, cisterns under intersections could cope with the sudden surges of storm runoff while offering water for firefighting and storing water for summer. Cisterns, wells and in-house energy-water systems translate to water and power remaining available even if the city were hit by a massive earthquake.

Los Angeles River restoration would start at the base of the hills and move downstream to a restored salt marsh between San Pedro and Long Beach. It would be a big construction project. Concrete would be removed and crushed into gravel to be used in the new riverbed. Landscape contractors would plant native species of grasses, scrubs and trees, then sit back and wait for the first storms to rearrange everything they'd just done.

A gutter of a river would become a grand marine boulevard. Riverside property values would skyrocket along with the growth of trees in suddenly valuable back yards. New pedestrian and vehicular bridges, like modern versions of Florentine bridges, could be lined with commercial space and broad

Systems for groundwater restoration and sewage management can be neighborhood-scaled. Permeable paving can result in grass where there is now asphalt, and the ability to capture rainwater. Cisterns at major intersections would provide surge tanks to cope with sudden storms and for fire-fighting. Reducing runoff would allow concrete drains to be restored as creeks. Sewage would be digested in ponds, providing a park and flowers.

sidewalks overlooking the long greening path of braided trees and shrubs. In winter it would be a wide river of brown, settling to a broad stream in spring and modest creek in summer.

Wildlife would return to the river. The combination of a riverbed alive with vegetation and lined with trees, plus backyards and many roofs partially covered in grass would no doubt cause a population explosion in butterflies and birds. Coyotes, whose presence is already begrudgingly accepted in the Hollywood Hills, would discover the river as a pathway from marsh to mountain. Swallows would nest in river bridges, darting amidst the trees at dusk. Raccoons and other small mammals would thrive. Pathways alongside the river and across the riverbed in summer would offer residents an experience apart from the city.

Inevitably the wild river and rising ocean would rejuvenate downstream marshes. Former oil refineries, as well as thousands of acres of related industrial properties, could become salt marshes. These estuaries would again become the marine nurseries they were, providing habitat for young fish, and a food source for innumerable shoreline species. The fishermen of San Pedro would notice a steady increase in the quantity and quality of all kinds of fish in the LA harbor, where once-naked rock seawalls would be carpeted in life, a change due in part to the conversion of ships to hydrogen-electric — ending pollution. Pacific species would also increase, as would seal and whale populations. Hollywood restaurants would feature *local* fish and shellfish.

The region's fleet of vehicles would be hydrogen-electric. No smog or fog. Crystal clear days would be common and residents would become accustomed to seeing snowcapped mountains above Pasadena. Office workers might occasionally be startled by condors cruising by the windows of downtown high-rises as they ride the warm updrafts. At night the city would sparkle, but streetlights would be much dimmer and seem to mirror the stars. Up in the hills, on a warm spring evening, residents out walking just might not notice a Hollywood star because they'd be staring at the real stars.

On May 10, 2007, the Los Angeles City Council voted 12-0 to restore all 31 miles of the Los Angeles River. The plan will involve over 200 small projects, including several new road and pedestrian bridges and the restoration of steelhead trout. It is expected to cost up to $2 billion over the next 15 years.

11

Blood, Water and Crystals

Vision is Health

WHETHER THE TALK is of power plants or power drinks, most people tend to focus on the hardware, not the software. They see the real tangible things, the visible power plant or the visible glass of a bottle, but not the software or recipe necessary to make it. Few people would even consider whether some malady they are experiencing was caused by their picture of the future and how they feel about it; they'd rather they'd seek a physical cause that related to the physical symptoms. It might be a bacteria or a virus, or perhaps a poison, but in any case it's a real physical entity attacking one's real physical body. But could it be that the most critical quality of our being, the quality that affects our health more than any other, is the seemingly simple belief that we have a future? Do we have a software problem?

While there is considerable argument about the specifics of climate change or the decline of oil or the rise of oceans the overbearing emotional reality is uncertainty. We cannot depend on energy prices, economic prosperity, global stability, ecological health or even the location of coastlines. A great many believe all they can expect is increasing uncertainty and a steady decline of civilization — a dead end.

We choose the future we are going to believe in. In facing a job interview we all assume our chances of success are best if we are positive. No one would advise another to glumly meet their potential employer while assuming they won't get the job, even though that would be a "realistic" assessment of one's odds. On the contrary, we are hopeful of a positive outcome because we know our behavior, our positive outlook despite all rational analysis of the odds, influences our chance of success. We choose to be "unrealistic."

As a species we have no widely accepted positive vision of our future. On the contrary, the popular vision of the future in many cultures, especially among young people, is a new brand of Armageddon. Instead of the standard variations on the gray police state theme we're now seeing Manhattan sunk with underwater views of fish swimming out of the Bleecker Street subway station. Presumably they are headed to a cafe for a poetry reading.

Most people throughout history assumed there would be a future for their children. Their sense of time was reinforced by the cycles of the seasons, of animal migrations, crops and the lives of those known and loved. It was expressed, if not enshrined, in the permanence of sculpture and architecture crafted by cultures that vanished thousands of years ago. Belief in a future is also expressed in more mundane realities. Agreements of all kinds, from marriages to car sales to corporate mergers to the sale of vacation property, are based on the essentially wishful assumption that all parties to the transaction will remain functioning, sane and capable for some period of time. Business deals are based on simple faith, not in legal language, but in the belief that buyers and sellers will be alive for at least a few days or weeks to conclude the deal; that their agents will be around for awhile in case someone made a mistake; and that the financial, legal and/or religious institutions involved will be solvent, trustworthy and ethical for some time to come. There are no life insurance salespeople visiting intensive care units.

Faith in the future is expressed by accountability in the present. A bank account is only real to the degree we trust the bank as an institution, which is why banks once carved the word "trust" into granite buildings that resembled the Parthenon. If it is rumored a bank is in financial trouble it's equivalent to saying the bank's future, and that of its depositors, is in doubt and therefore they cannot be trusted. If too many depositors lose faith and attempt to withdraw their money they can bankrupt the bank, thus destroying its future and damaging their own. A rumor can become a self-fulfilling prophecy.

If a group of people adopts the view there will be no future their motivation to enter into agreements with others all but vanishes, along with any notion of accountability. They needn't feel responsible to anyone if they believe they will vanish in the coming fire. If too many people adopt such a grim view of life's prospects we face a collective depression, in psychological and economic terms. *That* can be the end of a civilization, for without collective faith in our future our agreements, our very intentions, have no lasting meaning and we are all mere individuals accountable to no one.

We are headed into new territory, where the future may be so uncertain long-term agreements become problematic. Already it is becoming difficult for many people in coastal areas and river floodplains to obtain insurance because of the uncertainty and severity of weather, a reality compounded by many people living in areas vulnerable to dangerous storms. Already we're seeing images in movies and on TV depicting how climate change will result in the loss of Florida and much of the coastal US, and how it is all but inevitable within a century, perhaps within 50 years. It's the prospect of Miami as an artificial island of storm-washed highrises far out at sea, or a civilization lost undersea like some modern vision of Atlantis. We cannot even be sure of where the coastlines will be in the future. At what point in the next 50 years does it become impossible to get a mortgage in Florida because no one is certain if there will be a Florida? Or will the sea only rise an inch and stop? What greater uncertainty could there be?

Cultural trends, notably grim movies of bleak futures, reveal little faith in the future. People may feel positive about their personal situation, but fearful about larger realities. There is a general sense of cognitive dissonance caused by the conflict between what little we can control and a world seemingly gone mad. We read the news about melting ice, we feel the weather getting warmer or we know by our own study that rising heat and falling oil reserves mean we are in deep trouble. Yet the media presents the story with detachment, as if discussing the cost or practical realities of saving the world was just the latest in a series of issues in the news, right after the "fight against traffic" and "war on terror." As yet there is no apparent recognition that climate change, and a whole raft of environmental issues from ocean acidity to drought to rising seas to dying ecosystems, are all of a piece and the issue is saving the planet's biotic communities, including us. A large portion of the world's most influential and innovative thinkers recognize the problems, but their perception, as well as the wealth of new and positive trends, are given minor consideration in mainstream mass media.

"The messenger did it." It's easy to blame the lack of any significant attempt to call attention to the issues surrounding climate change, and many other pressing matters, on media ownership and editorial control. "Public information" campaigns emphasizing an obvious bias towards a specific vested interest, most notably the long-running oil company campaign designed to convince Americans climate change isn't happening, are often cited as evidence of how media companies manage opinion. There's no question media is influenced by powerful interests, if for no other reason than the owners of the media empire are buddies with the rich and powerful. But to assume they control all media is to grant them their hubris while demeaning the intelligence of everyone else — the audience.

We all seek information of immediate relevance, that can be trusted and is presented in a realistic yet hopeful manner. Media, whether news or entertainment, is not merely about the story, but about the culture. How stories are told and editorials written reinforces our vision of our community's future and that of our entire culture. Millions of people, worldwide, have deserted mainstream media in favor of the Net, in part because traditional media no longer represent *their* community.

No doubt many people do respond to the realities of climate change and other global problems with a shrug, followed by a blanket statement that civilization is over because people are "too stupid" or "too self-absorbed" to make a difference and save the day. But to assert such views today, in the face of profound environmental, social and economic problems, and the extraordinary implications of the Net, is to ignore what may be the first common experience shared by all humanity, the grand experiment of vision by global consensus happening right now. You don't need cash or a credit card to participate.

Some several million people are engaged in addressing these seemingly intractable and huge problems, but they are not on the nightly news. They are on the Net. They may be advancing their own vision or building upon the visions of others, but they are all focused on a vision. Random searches using terms like "solar-energy," "biodegradable plastics," "permaculture farming," "fuel-cells" or "hydrogen production" will yield innumerable hits, for individuals and companies and agencies, all collectively defining a revolution. It is a revolution growing up on its own, and it is not based on realistic analysis but unrealistic, hopeful and often outrageously wishful thinking. If there is anything all participants share it's an unbounded sense that we, all people, can do anything, and that vision is the only antidote to terminal hopelessness. It's a matter of faith.

Faith is not merely a religious matter, but an extremely pragmatic reality that both reflects and influences our lives. When drugs are tested on a random selection of people, half are given the real drug and half are given a sugar pill — a placebo. A significant portion of those taking the placebo report the effects of the drug even though nothing physically happened. Ideally, presuming the drug works at all, a majority of those who took the real drug report the expected results. Ostensibly this proves the drug's efficacy. But such a study is in part based on the belief shared by most, if not all, participants that taking any medical drug will positively affect their health. Thus while the study may prove some measure of success there remains an element of faith at play, not merely the certainty of biological or chemical functionings. Taking a drug represents faith in the drug's ability to *alter* the future, and this expectation of change is present whether one takes the real thing or the placebo. In either case there is a measure of faith involved, but who can determine the line where the effects of faith end and cold hard biological reality takes over?

Talking doom in a crisis can be tantamount to criminal behavior. Soldiers hunkered down in a bomb crater and faced with a dawn battle against a superior force, knowing losses will be heavy, do not want to discuss lousy odds. Such talk is an invitation to get killed. In such a context being realistic means defining a strategy to maximize everyone's chances of survival based on an honest assessment of the odds. Any suggestion of losing is simply unacceptable.

Which is better for our health, "realistic" assessments of the risks based on someone else's opinion taken out of context, or a faithful assessment of one's potential at that moment based on what everyone present in that moment *wants* to happen? In the foxhole of life we're never truly ready, there's never a helpful consultant at our side to provide salient advice as we aim, and everyone is always making an educated guess when they fire. We can only assess our immediate resources, collectively witness our faith — an expression that may be nothing more than a knowing glance between friends — and then go forth out of our past with an ever larger vision. A vision that embraces realistic assessments, yet is focused on the best in us, and on what we feel as much as what we know.

Since the sixties, millions of people have turned off the television and tuned in their own channel. Different people found different paths, for some it was politics, others science, art or religion, still others explored psychology, mythology, medicine, law or design. They were political radicals of all stripes; beatniks seeking a new American literary voice; hippies spawning a new spiritual vision; counterculture intellects wrestling with global issues; scientists bent on defining a more holistic vision of nature; artists who found their vision in a larger spirit of humanity; feminists who built a new consciousness of women's rights; Native Americans who rediscovered their religion; environmentalists who envisioned a gentler way of life; craftspeople who found many old ways were still good ways; farmers who rediscovered the economy and joy of organic farming; technologists who saw power in little technologies big companies ignored; and health practitioners who found their power in subtle energies known to the ancients. All the millions of people represented by these categories, encompassing an extremely disparate portion of the world's population and largely concentrated in the bigger cities of the world, have perhaps only one quality in common. They all came to recognize how their personal vision of their life was the source of their power and health. Growing from this realization, they gradually forgot how to speak the official company story as defined by any authority, and started to live their own story.

This profound need to explore one's vision is central to the concepts of liberty and equality, concepts that give license to become one's own best work of art without fear of constraint by state, society or church. It is naturally human to want to go for the stars, reach for our dreams and follow that yellow brick road to be the best we can be. It is naturally human to tell a story of today, yet highlight one's hopes for the future. It is also naturally human to want attention for our creativity and perhaps feel part of some larger reality.

The revolutionary nature of the trends that began in the sixties, and have since broadened and been elaborated on the Net, is still not grasped by mainstream media, or what is fast becoming the backwater media. Individual reporters and some publications grasp pieces of it, but not the whole pie, nor the implications of all of it at the same time. It is a new society growing from within the old. It isn't based on the notion that institutions manage the world, individuals have no power and everyone is motivated solely by

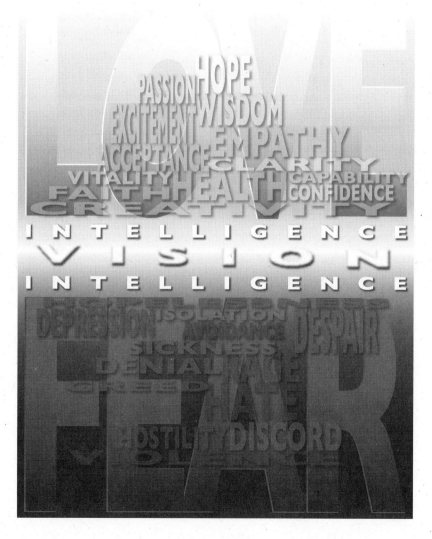

Health is usually discussed as if it were simply a matter of physical realities. In complex industrial societies, and in developing regions, the stress of change, lost livelihoods and social dislocation can lead to violence and despair. The only antidote is the development of vision, a story of one's future defined by the realities of the present and knowledge of the past. What would be the value, in health and the well-being of families, communities and all humanity, of a shared vision that was positive, attainable and immediately possible?

the most elementary biological needs and the most banal of interests. This new society, as displayed on innumerable Websites and in countless books and other media, is based on honest assessments of the realities we face, acknowledgements of the seeming impossibility of it all, yet a hopeful vision of what to do about it.

Could it be that the most healthful and helpful response to climate change and a raft of other issues has been evolving for three decades, but because old media inherently divides reality into a mincemeat of issues and ideas few have noticed the trend? What would be the effect on our general state of physical and emotional health if some significant portion of the population possessed a clear, practical and eminently attainable vision of change, of a world where environmental restoration, social equity, civil rights, education and economic development were woven into a new future? Would therapists notice a decline in patients? Conjure a vision and call in the morning.

One day the Soviet Union ended. Everyone already had another vision. It was time to do something else.

Pathology of Technology

Amazingly, just as the implications of a pervasive sense of doom on our health is largely ignored so too is the health of the environment ignored as a precondition for our health. Emphasis is placed on the individual and their symptoms, sometimes their nutrition, but not the water they drink, the air they breathe and the earth they stand on.

There can be questions about the specific results of innumerable chemicals, hormones and heavy metals in our bodies, but there can be no question this synthetic soup is having effects on the health of people, animals and plants. No matter how well studied, the impacts of any one chemical on all life cannot be known.

Each one of us is a walking ecosystem with a thin boundary between what is "I" and what is "they." Between our skins there is a soup of gases called the atmosphere and it contains a mob of microbes cruising along in the breeze, buffeted by the winds of our breath and often clinging to soot or dust like survivors on a raft. These tiny critters move around us, on us and through us as if we were little more than a glob of foam.

Virtually all of us are now carrying substances we did not choose to consume. We ate food laced with pesticides or we worked in a place where poisons were present or we simply lived in cities where the air was laden with pollution. In the end we and the animals and the plants are left with the consequences of actions taken unwittingly by ourselves, our predecessors or people a long time ago we cannot even name.

Health and personal security are imperatives of our immediate future. We need to respond to global warming by reducing greenhouse gas production and we need to reduce the pollutants being put into the waters while

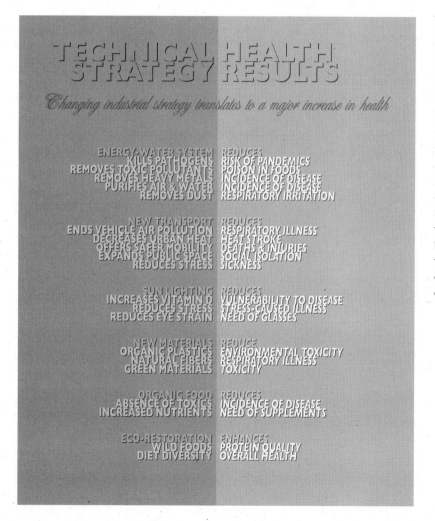

TECHNICAL HEALTH STRATEGY RESULTS

Changing industrial strategy translates to a major increase in health

ENERGY-WATER SYSTEM	REDUCES
KILLS PATHOGENS	RISK OF PANDEMICS
REMOVES TOXIC POLLUTANTS	POISON IN FOODS
REMOVES HEAVY METALS	INCIDENCE OF DISEASE
PURIFIES AIR & WATER	INCIDENCE OF DISEASE
REMOVES DUST	RESPIRATORY IRRITATION

NEW TRANSPORT	REDUCES
ENDS VEHICLE AIR POLLUTION	RESPIRATORY ILLNESS
DECREASES URBAN HEAT	HEAT STROKE
OFFERS SAFER MOBILITY	DEATHS & INJURIES
EXPANDS PUBLIC SPACE	SOCIAL ISOLATION
REDUCES STRESS	SICKNESS

SUN LIGHTING	REDUCES
INCREASES VITAMIN D	VULNERABILITY TO DISEASE
REDUCES STRESS	STRESS-CAUSED ILLNESS
REDUCES EYE STRAIN	NEED OF GLASSES

NEW MATERIALS	REDUCE
ORGANIC PLASTICS	ENVIRONMENTAL TOXICITY
NATURAL FIBERS	RESPIRATORY ILLNESS
GREEN MATERIALS	TOXICITY

ORGANIC FOOD	REDUCES
ABSENCE OF TOXICS	INCIDENCE OF DISEASE
INCREASED NUTRIENTS	NEED OF SUPPLEMENTS

ECO-RESTORATION	ENHANCES
WILD FOODS	PROTEIN QUALITY
DIET DIVERSITY	OVERALL HEALTH

Renewable energy, new transportation initiatives, sun lighting, green materials, organic farming and eco-restoration all represent a new industrial strategy. If fully developed this strategy would not only respond to the major problem of global warming, but also address a raft of health problems. Unlike medical approaches, where drugs, surgery or other treatments focus on the individual, these strategies represent the healing of the social body and the planet by removing sources of disease, injury and premature death.

removing pollutants already in them. Ending this pollution would seem to be the most important investment we could make for the health of all species. We must end the production of inorganic chemicals freely used in the environment.

This may also be the first time an investment in health would also be an investment in defense. Given that anyone can possess a weapon of mass destruction, in the form of a vial dropped in the waters, and given that all heavy weapons are useless in the face of such threats, it would seem critical that we develop an energy-water system that by its very functioning would eliminate everyone's vulnerability to poisons spread via the water. Cracking water electrolytically is perhaps the most certain means of purifying water and concentrating pollutants, simply because the water becomes its two constituents and nothing else. Only oxygen and hydrogen molecules pass through the membranes of electrolyzers and fuel-cells.

In addition to the horrors of intentionally introduced microbes we are also faced with diseases for which we have little or no resistance. Some diseases once demolished by antibiotics are now resistant, notably a form of staphylococcus that apparently evolved in hospitals. Diseases for which we have no resistance and no known cure can often be easily transmitted by water, and rising temperatures, rising oceans and melting tundra are a recipe for the growth of microbes.

Those in developed nations live in a world linked by incoming fresh water pipes and outgoing sewage pipes. Most municipal water systems are managed by folks who know their waters. But they are often faced with tight budgets and major reconstruction costs. Generally sewage treatment plants are comparatively new, often dating from the 1970s or 80s, but in many cities the actual sewer pipes are often in need of serious repair. Water and sewer systems can carry infectious substances whether introduced to the system by intention or accident. Pipes are a microbe highway.

We are meant to drink pure water. US federal standards for water purity have "allowable" quantities of cyanide, radium, uranium, mercury, arsenic, lead, zinc, fluoride, hexavalent chromium and several other zesty flavors. These quantities are tiny and some contaminants are natural, but many are not. We may need some small amount of the naturally occurring ones, but in excessive quantities they are deadly. Those that are unnatural represent molecular structures our bodies have no mechanism to cope with. The very existence of unprecedented foreign chemicals carries profound meaning as scientists delve further into the mysteries and subtleties of genetics and cellular chemistry, a realm of relationships at once both elegant and robust in its functioning, but vulnerable to interference by the tiniest bit of some chemical compound. Yet it's widely assumed that through solid science the Environmental Protection Agency knows these tiny quantities are harmless. Really?

Development of a new energy-water system would break the continuity and radically reduce the extent of water and sewer systems while purifying water in each building or complex of buildings. You could drink pure water. It would be beyond federal standards, it would be clean. You could turn on the under-sink water enhancer and drink "custom" water by doctor's prescription. It could be precisely remineralized for optimum balance of minerals and electrolytes, and ionically enhanced with mild alkalinity to balance body acid. Water would be an agent of health, not a source of uncertainty.

Fleets of Micro-Machines

Both ecological restoration and our health require removing poisons in the environment, especially the waters. In proportion to the world's oceans, the bays and estuaries affected by urban waste are small, but they are the most critical habitat for breeding fish. The coasts off hundreds of the world's cities are toxic waste dumps of household garbage, industrial odds and ends, drums, cans and bottles of god knows what. Even subway cars have been dumped into the sea — fish on the "A Train."

These repositories of civilized shit are little more than heaps of decaying garbage coated with silt that rains down from currents above, silt often bearing heavy metals and inorganic pollutants, such as pesticides from lawns or solvents from a factory.

Pollutants wafting off the trash dumps, and the heavy metals and pesticides layered in the silt, can kill or severely diminish the potential of any ecosystem.

Given the state of these dumps it would be insane to physically remove these materials by dredging. Shovel and vacuum dredges are inherently inaccurate machines, and the size of such machines precludes their use in backwaters and narrow channels. Moreover, dredging risks disturbing delicate materials or breaking open containers with substances that would then be released to the waters. But it could be possible to remove the offending substances with far less disruption.

It is possible to build very small machines. Consider a submarine the size of a minnow, with a photovoltaic battery, camera and computer. It would surface to recharge. It would be programmed to seek out substances, such as a specific chemical or even radioactive particle. It could burrow into the silt

While often discussed as medical devices, micro-machines could also be used to clean up pollution, thus reducing a cause of many diseases. This illustration shows new technologies used in a tiny submarine, analogous to a tube with rotating teeth at both ends. The sub would float on the surface until its photovoltaic skin charged its batteries, then it would submerge and burrow into the silt searching out microscopic quantities of pollutants, which it would stow in a tiny storage tank. Once filled it would return to its starting point, where millions of the tiny machines would be emptied and then returned to service.

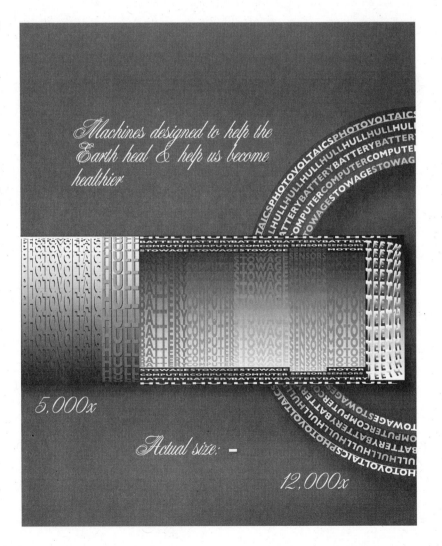

Machines designed to help the Earth heal & help us become healthier

5,000x

Actual size: —

12,000x

and seek out mercury. Once it filled its on-board tanks it would return to a central collection point. If a shark ate one it would just pass through, perhaps producing a Discovery Channel documentary in the process.

Unlike chemical or genetic techniques tiny machines don't pollute and don't replicate. They could be designed and programmed to do a specific set of tasks. Invasive species, especially plants, are a particularly vexing problem because they can quickly become pervasive on a very large scale and thus present a huge cost to eliminate them. Some alien plants and animals find a niche in a new environment and don't bother their new neighbors, but others, such as kudzu in the Southeastern US or water hyacinth in California, have no predators so they grow wild and strangle their neighbors. This usually results in a far less productive environment.

A fleet of subs might seek and destroy a critical part of the hyacinth's root structure, thus setting the stage for a massive expansion of native reeds and related species. A fleet of millions could be programmed to kill a species of marine grass now strangling Mediterranean fisheries, and yet another could assist in seeding the sea floor with native species.

Micro-machines are being conceived for internal surgery — swallow this scalpel please. The same technology would be viable for wide range of restoration tasks, on land and in the water, where the scope and complexity of the problems are daunting and only a large-scale and simultaneous response has any hope of success.

Many micro-machines would do their job, wear out and stop working. They could be made largely of silicon. Silicon is made from quartz, which comes from common beach sand. The little machines would become sand again.

PART FIVE

Wealth By All Measures

12

Gold in the Sun

Profit As If Everything Matters

PROFIT IS SURVIVAL. Profit is a measure of the gain in value between the energy cost of taking an action and the benefit derived from that action. If the squirrels don't store more than enough acorns for the winter they might just starve — they probably do ten percent net acorn profit. People equate profit with money, but money is not the only measure of profit, just the only one commonly measured. Everyone seeks profits of some measure and to varying degrees; some seek monetary profits, while others seek a quality of life profit in improved health, creative work or social opportunity. In the future it will be better — better is profit.

Automakers and all kinds of consumer product companies increasingly measure success by a much broader measure than monetary profits only, such as increasing market share on a new product that represents potential profits, lowering costs by reducing pollution, or by achieving higher customer satisfaction ratings. They recognize monetary profits may be elusive in some years, often due to forces beyond their control, and during those periods their long-term survival requires the support of a community of customers, shareholders and stakeholders who believe in them, people who may have no facts to rely on except measures of the company's potential and promise.

Oil, coal and utility companies have all but forgotten other profits. The energy business is focused on money and price. Other measures barely matter when you produce a commodity exactly like the commodity produced by your competitors. No one can fundamentally change gasoline. If you can't change the product and charge more because of the valued added, then the only means to sustain and maybe increase profits is to reduce costs, reduce costs and then reduce costs some more, all while keeping a wary eye on commodity markets that can go crazy over anything from civil unrest in Saudi Arabia to car sales in China.

The money-driven perception of energy companies blinds much of the industry to the trends that drive the development of new infrastructure, trends that respond to issues beyond the purview of existing companies, that involve new technologies offering a higher quality of energy, and that define new values and new prices. New infrastructure only gets built when the majority of customers perceive it as profitable to them by every measure — economic, ecological and cultural. New energy and water technologies meet that standard because they translate to a dramatic reduction in resource use, an extraordinary increase in personal security, a massive reduction in distribution infrastructure, an end to air and water pollution, and possibly the amelioration of climate change. They offer the greatest benefits to the largest possible population in the least amount of time. But the energy business is only looking at cost, price and profits right now, not the values of new infrastructure that commands a higher price because it does more.

The Whole is Worth More than the Sum of its Parts

The new energy and water technologies change those businesses. Instead of selling just energy or just water one is selling the *means* of producing energy and water. Instead of producing gasoline, diesel fuel, natural gas, propane, oil, coal and uranium, all for different purposes, it's possible to produce one system for virtually all needs. Instead of custom facilities taking years to develop, new energy-water technologies allow development to occur all over the place. Instead of obtaining, processing and distributing energy and water from elsewhere it becomes possible to use the energy and water delivered by Universe Inc.

Sunlight, wind and rain are equitably priced commodities — they're free. The profit arises from making these natural commodities useful. Compared to continually trying to find profits in selling commodities, new energy and water technologies offer at least the potential of higher financial profits because they represent superior values. Photovoltaic shingles are not just generating electricity, they're a long-lasting roof. A fuel-cell provides not only electricity, but also pure water. By addressing multiple needs each component increases the overall value of the system, so prices are determined not just by the cost of a commodity, but by the value provided to customers.

This hypothetical integrated energy-water appliance, a water and power company in a box, is the key product. It would be a generic standard manufactured in many forms by many companies. A home system might arrive in two boxes. Inside one would be a refrigerator-like unit containing the electrolyzer and fuel-cell, as well as filters and controls; the second would include tanks, photovoltaic panels, wires and hoses, and several bags of nuts and bolts. There would be standard models, and custom versions for larger buildings and factories. One visit to a manufacturer's Website, one phone call — and one water-energy system delivered anywhere.

This hypothetical appliance represents a very large potential industry. Unlike centralized power plants, dams, grids and reservoirs, all of which involve major politics, an energy-water system in a box involves little or no such complexities. In the developing world this means the infrastructure can grow from the grassroots and be scaled up over time, just as the photovoltaic business is now developing.

In the first decade of the 21st Century the world's utility industry is trying everything. There's a contingent that wants bigger power plants; another focused on conservation; and another, is on renewable energy. Meanwhile hundreds of conventional power plants are being built to add capacity to national grids. Others have made up their minds: Pacific Gas and Electric Company, a San Francisco-based utility, began running a television commercial in 2006 that says: "The Future is Renewables."

Obviously changing infrastructure is difficult. Changing it within 20 years would seem impossible in a world where it may take longer to write the environmental impact statement than to build the power plant. Yet consider that in the four to six years required to design and build one large power plant, the thousands of buildings it would electrify could be re-roofed with photovoltaics, well before the power plant was even finished. There would be no environmental impact statement because there wasn't any, and no political opposition because there was no impact.

Critics of hydrogen fuel-cells and renewable energy see high prices and invariably argue that it will be decades before most cars are commonly run by hydrogen. They assume that a product not competitively priced now is useless. But what matters are not just the contemporary prices, but whether the inherent economies of these new technologies will result in competitively priced products. In 2007 photovoltaics and fuel-cells are too expensive if compared to established technologies, but current research and new products about to be introduced clearly reveal the potential of competitive,

In a world of specialists, each focused on one or another idea or technology, it's difficult to discern the larger value of any one idea, let alone the larger values of many ideas. Nevertheless, given the complexity of issues we face it would seem imperative that we view every concept inclusive of the broadest range of impacts. Further, that we seek ideas and technologies which are inherently complementary, offering the greatest possible results with the least action and resources.

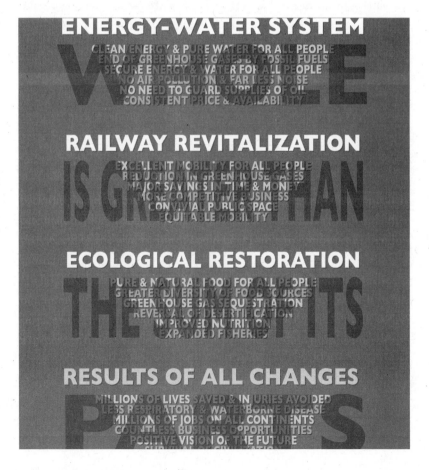

ENERGY-WATER SYSTEM
CLEAN ENERGY & PURE WATER FOR ALL PEOPLE
END OF GREENHOUSE GASES BY FOSSIL FUELS
SECURE ENERGY & WATER FOR ALL PEOPLE
NO AIR POLLUTION & FAR LESS NOISE
NO NEED TO GUARD SUPPLIES OF OIL
CONSISTENT PRICE & AVAILABILITY

RAILWAY REVITALIZATION
EXCELLENT MOBILITY FOR ALL PEOPLE
REDUCTION IN GREENHOUSE GASES
MAJOR SAVINGS IN TIME & MONEY
MORE COMPETITIVE BUSINESS
CONVIVIAL PUBLIC SPACE
EQUITABLE MOBILITY

ECOLOGICAL RESTORATION
PURE & NATURAL FOOD FOR ALL PEOPLE
GREATER DIVERSITY OF FOOD SOURCES
GREENHOUSE GAS SEQUESTRATION
REVERSAL OF DESERTIFICATION
IMPROVED NUTRITION
EXPANDED FISHERIES

RESULTS OF ALL CHANGES
MILLIONS OF LIVES SAVED & INJURIES AVOIDED
LESS RESPIRATORY & WATERBORNE DISEASE
MILLIONS OF JOBS ON ALL CONTINENTS
COUNTLESS BUSINESS OPPORTUNITIES
POSITIVE VISION OF THE FUTURE

but not necessarily lower prices, within a few years for photovoltaics and three to five years for fuel-cells. Both technologies inherently achieve efficiencies by using the least resources, specifically thin films and membranes. This means the design of miniscule structures made of atoms tends to be the defining factor in efficiency and cost, not the volume of resources nor inherent complexity of manufacture.

New technologies change the price equation. Ten years from now energy and water prices might be higher. But vehicles and homes would need less than a quarter the energy and a fraction the water, so the utility bill might be the same or lower. Predicting future prices, given the array of uncertainties we face, is pure speculation. We can only assume oil prices will likely rise and renewable energy prices will decline. Both are moving targets.

New values are the wild card. What is the value of an energy and water system that would power your car and home, cause no pollution and provide pure drinking water; a system not vulnerable to blackouts or energy shortages; a system that doesn't require a vast investment in hardware, nor a military presence anywhere on earth; a system that by its use addresses global warming and purifies the water; and a system that uses the natural distribution systems commonly known as sunlight and rain?

New energy and water technologies, especially if marketed as household appliances, would greatly facilitate a rapid transition. They can be mass produced and installed by electricians and plumbers. Installing PV roofing and all the components of a hydrogen-based energy and water system could be done by local contractors in a week. Another week for new windows and skylights. Transforming commercial structures might take a few months, a high-rise six months to a year. A whole city, such as Los Angeles, could be done in 15 years. There could even be a TV show on the entire event of transforming LA, a city makeover.

We Are Already Spending the Money

Virtually all the technology needed to create a new energy and water system exists, with many components already in mass production. Photovoltaics are produced by several companies, as are all the necessary electrolyzers, fuel-cells, hydrogen tanks, water filters, air cleaners, carbon scrubbers, inverters, valves, controls, wire, pipe, hose, screws and nuts. New software would be required, and the entire system would need to be designed as a unit. Technical feasibility is not at issue. Economic feasibility is a matter of choice.

Contrary to popular belief infrastructure revolutions do not proceed with rational planning, all technologies proven and all funds in the bank. Major infrastructure revolutions have started with insufficient information in unexpected places at the worst time. The transcontinental railroad was built *during* the Civil War, and many of the biggest dams in the western US were built *during* the Depression. These events were driven by political, cultural, economic and religious values that eclipsed wars and bankruptcies.

Contrary to popular belief, infrastructure is not hugely expensive. The world is made of things that cost billions of dollars, from $13 billion for an aircraft carrier to $8 billion for an offshore drilling rig, $5 billion for a major freeway interchange and $2 billion for a measly highway bridge. Many corporations make more than $13 billion just in profits. Large numbers don't matter, but the cost per person or per mile or per ton or per house does. If the cost per unit is lower than existing costs we're saving money, if it's higher we're spending more. A higher price raises the question: what are we gaining? In either case the big numbers to build a railway may seem impressive, but the ticket price is what matters.

Buyers won't buy hydrogen electric cars unless there are hydrogen gas stations and there will not be hydrogen gas stations if there aren't any buyers — a chicken and egg situation. Critics of hydrogen often cite the "enormous" cost of outfitting a quarter of the 100,000 gas stations in the US with hydrogen facilities — $13 billion according to some estimates — as a problem. In the realm of infrastructure $13 billion is pocket change.

The auto and oil industry, through the California Fuel Cell Partnership, is already doing the same thing with hydrogen they did with diesel fuel in the 1970s: focusing on vehicle fleets. Fleet owners, such as transit districts, build the hydrogen stations and private buyers initially use their system until there are enough private buyers to justify separate stations. Government grants and tax incentives are valuable, but most of the new infrastructure can be developed using money we would spend anyway.

Several countries, states and cities, as well as many corporations are embracing new energy and water infrastructure, either as policy shift or new business opportunity. Photovoltaic systems and wind farms are proliferating in Europe, especially Germany, as well as in Japan and throughout Asia, Australia and Africa. In the US new energy and water technologies are showing up in buildings from the San Jose headquarters of Adobe Systems and Google, to the new Bank of America tower in Manhattan, to the new California Department of Transportation offices in Los Angeles, where the south-facing side of the building is almost entirely covered in photovoltaics. Many new buildings now collect rainwater, dramatically reducing their reliance on external sources. It would appear the world has made the choice, given the sheer number of companies, nonprofits and government agencies involved in a diversity of activities related to new infrastructure.

ENERGY-WATER SYSTEM APPLIED TO ALL HOMES IN THE CITY & COUNTY OF LOS ANGELES

TOTAL POPULATION	10,300,000
TOTAL AREA	4,034 SQUARE MILES
TOTAL HOMES	1,300,000
AVERAGE PRECIPITATION	15 INCHES
GALLONS OF RAINWATER/YEAR	1,060,000,000,000
GALLONS OF WATER CONSUMED/YEAR	507,000,000,000
AVERAGE INCOME PER CAPITA	$22,700
AVERAGE ELECTRIC BILL P/MONTH	$115.00 ('07)
AVG BILL PER CAPITA PER MONTH	$14.58 ('07)
AVG PRICE PER KILOWATT HOUR	$.13 ('07)
TOTAL CONVERSION CAPITAL	$24,000,000,000
POTENTIAL WATER USE	202,000,000,000

YEAR	1	2	3	4	5	6	7	8	9	10	11	12	13	14	15
HOMES CONVERTED	2K	3	4	13	26	52	104	117	130	143	143	156	156	143	104
RESIDENTS SERVED	20K	51	102	205	412	823	1.6M	2.5	3.6	4.7	5.8	7.1	8.3	9.5	10.3
JOBS/INSTALLATION	0.6K	0.8	1.4	2.9	5.8	11.6	23.3	26.2	29.1	32.0	32.0	34.9	34.9	32.0	23.3
COST PER HOME	$25K	$24	$24	$23	$21	$19	$18	$18	$18	$18	$17	$17	$17	$17	$17
CAPITAL ALL HOMES	$65M	$96	$158	$302	$550	$991	$1.8B	$2.0	$2.3	$2.5	$2.4	$2.7	$2.6	$2.4	$1.7
CAPITAL PER CAPITA	$3.1K	$3.1	$3.1	$3.0	$2.8	$2.6	$2.4	$2.3	$2.3	$2.3	$2.2	$2.2	$2.2	$2.2	$2.2
AVG BILL PER MONTH	$187	$185	$181	$174	$158	$142	$135	$134	$133	$131	$130	$129	$129	$129	$129
AVG BILL PER CAPITA	$19	$19	$18	$18	$16	$14	$14	$13	$13	$13	$13	$13	$13	$13	$13
AVG COST PER KWH	$.35	$.35	$.34	$.33	$.32	$.29	$.27	$.26	$.26	$.26	$.25	$.25	$.25	$.25	$.24
PRCT OF INCOME	1.0%	1.0	1.0	0.9	0.9	0.8	0.7	0.7	0.7	0.7	0.7	0.7	0.7	0.7	0.7
EXISTING KWH	13.8T	13.8	13.7	13.5	13.2	12.7	11.6	10.3	9.0	7.5	5.9	4.2	2.6	1.1	0.0
PRCT OF EXISTING	0.12%	0.3	0.6	1.2	2.4	4.8	9.6	15	21	27	34	41	48	55	60
PRCT OF WATER USE	0.08%	0.2	0.4	0.8	1.6	3.2	6.4	10	14	18.4	22.8	27.6	32.4	36.8	40.0
PRCT OF RAIN/YEAR	0.03%	0.10	0.19	0.38	0.76	1.52	3.05	4.8	6.7	8.7	10.8	13.2	15.4	17.5	19.0

Row notes:
- YEAR: Following a one-time design stage
- HOMES CONVERTED: Number of homes converted to 100% solar-hydrogen energy-water system in year
- RESIDENTS SERVED: Number of people served by systems developed, including children
- JOBS/INSTALLATION: Thousands of people employed in conversion of homes
- COST PER HOME: Thousands of dollars to convert the average home over two weeks, declining by innovation & economies of manufacturing
- CAPITAL ALL HOMES: Capital invested by utility company in buying new systems, per year, in millions & billions
- CAPITAL PER CAPITA: Capital invested expressed in thousands of dollars per person
- AVG BILL PER MONTH: Average monthly bill would remain the same as a fixed finance charge, no matter how much power used
- AVG BILL PER CAPITA: The amount per month if every person paid their own electric bill
- AVG COST PER KWH: Comparable cost based on the price of the system with no value assigned to water and heat provided
- PRCT OF INCOME: The cost of this system would be insignificant compared to average per capita income
- EXISTING KWH: As new systems were developed trillions of kilowatt hours produced by natural gas, coal, hydro and nuclear would end
- PRCT OF EXISTING: Using available conservation strategies the new system would reduce total need for electricity
- PRCT OF WATER USE: As systems developed water demand would fall to less than half pre-conversion water consumption
- PRCT OF RAIN/YEAR: Residents would use a fifth of annual rainfall, commercial would use less, & much of the rest would return to rivers and ground

The economics of developing the proposed energy-water system in Los Angeles, with all houses converted in 15 years. Innovations now happening suggest a target price of $25,000 for an average home conversion is attainable. The system would include photovoltaics, in some cases wind turbines. While the cost per kilowatt hour would be higher, demand would be lower due to more efficient appliances. The system would produce heat and pure water, and remove greenhouse gases, values not counted here. Commercial properties would be converted in parallel. Capacity to power a car would be additional.

Follow the money. Investment in renewable energy has been growing steadily for 20 years. Until the late 1990s the oil industry dominated photovoltaics, while a handful of small companies in wind turbines and solar heating grew slowly. BP is now the only oil company with a major presence in photovoltaics. Shell is engaged in research on new PVs and may yet be a major player, while Chevron-Texaco is involved in hydrogen research. Industrial consortiums like Sharp and Siemans are now the primary players in photovoltaics, along with new firms like Sunpower, Nanosolar and Miasole. GM and Chevron are involved with a company called Energy Conversion Devices in hydrogen and photovoltaics. Ford, Toyota, Honda, Daimler-Chrysler, Hyundai, Nissan and several other companies all have fuel-cell electric cars, with 140 operating in California in 2006.

Renewable energy and green technologies weren't especially attractive in the early years of this century. Then in 2005 and 2006 interest in green investments exploded. Major investors now include Swiss Re, a highly respected European insurance company. Curiously, from 1905 to 1920, almost exactly one hundred years ago there was an explosion of investment in electric trains, automobiles and gasoline.

Follow the water. Many city water agencies are developing conservation, recycling and groundwater restoration. Organic farming is growing, as are natural food stores. Many major corporations are manufacturing new water filtration and purification technologies, as well as appliances that use less water. A wide variety of organizations, including nonprofits, government agencies, ranchers, utility companies and real estate developers are engaged in ecological restoration, often motivated by the need to save water, purify water or restore the local waters; perhaps a creek, wetland or shoreline.

Follow thousands of students attending classes in high-tech, nano-tech, bio-tech and eco-tech in dozens of schools where the bizarre realm of quantum physics collides with the rigorous realities of initial public offerings. In this world future technologies are so subtle in function they are likened to a living creature. A casual perusal of the Web, a veritable model of organic culture, will reveal a dizzying array of papers growing from the convergence of the technological, biological and ecological.

If we followed new water and energy tech to the source of political power behind it all, we would find nothing. There is no central authority commanding the energy army. While it's fashionable to claim the ubiquitous "they" would oppose sweeping change in energy and water because utilities can't own the sun or automakers don't like electric cars, such assertions are based on the idea someone has a master plan. The management of several dozen automakers, oil companies, utility companies and associated banks do represent a *de facto* club of influential citizens, who often do belong to various real clubs with real agendas. But this "they," so often blamed for the world's ills, don't necessarily have any better grip on our future than any other "they," nor have anywhere near the influence others think they have upon the unfolding future.

Corporations are equivalent to committees with personal characteristics. There are dim corporations, brilliant corporations, terminally boring corporations, boringly normal corporations, incredibly vacuous corporations, breathtakingly competent corporations, arrogant corporations, psychopathic corporations and just the nicest corporations you'd ever want to meet. There are liberal corporations where anything is cool, and conservative corporations where everything is cold. There are corporations who embrace new ideas and ride the cutting edge of culture. There are even corporations who loathe those who bring up new ideas, but then they usually do this just before they die.

Then there are corporations that take bold steps. Companies like Sharp, BP, Sony, Honda and General Motors with their pioneering work in hydrogen-electric cars. Over the last 15 to 25 years these companies and many others have spent billions on advanced research with no reason to think the work would pay off anytime soon, if ever. Tens of thousands of people have been involved yet few outside the industry are even aware of what they have done or the staggering implications if they succeed.

The "they" often characterized as suppressing renewable technology and related innovations have been financing new energy and water technologies for some time, because it is or, they hope, will be profitable. "They" are not in business to lose.

Profit by All Measures

Renewable energy and related industries are about where computers were in the early 1980s, and the Net ten years later — poised to explode. Photovoltaics are available in several forms, at all scales. Wind turbines have become very competitive, from 6 feet to 140 feet tall. Stationary fuel-cells are competitive now, and automotive versions are likely to become competitive within five to ten years. Water filtration technologies are advancing, often making microscopic breakthroughs with macroscopic implications. Hydrogen technology is advancing on all fronts: production, storage and transport. All these technologies portend an industrial revolution of staggering magnitude.

High-tech journalists are forever seeking the stunning new invention that presages yet another high-tech revolution leading to round upon round of new investment — the "next big thing." By comparison to a new energy-water system, a potential industry designed with semiconductors and powered by semiconductors, any other high-tech thing pales by comparison. Not everyone wants to download the latest song but virtually everyone wants electricity to download that song and needs pure water if they're going to dance for long. Everyone.

New infrastructure is not happening in the old ways. No massive single investment need be made, and no centralized authority need plan the system. It's already growing by that organic silicon-driven creation called the Web. We can vote to raise taxes to fund technological research, buy railway routes or undertake ecological restoration, but tax funds would ultimately be a small component of total investment. Utilities, oil companies, automakers and hundreds of other companies, plus innumerable new companies, can finance the transition, and we'll just keep paying their bills.

Regional utilities would be the likely companies to manage the transition, install building systems and perhaps own or lease systems. Instead of buying a few big power plants they could use their financial leverage to purchase lots of home-scale energy-water systems. The utility or a local contractor would install a system. Some households might choose the optional auto-recharging system, at higher cost, and buy a hydrogen-electric car at the same time. The utility would maintain the unit and modify it from time to time.

Utility companies could undergo dramatic change. As corporations they could achieve higher financial and environmental returns on their investment, and their revenue would increase significantly since they'd be selling systems providing energy and water for buildings and vehicles. They'd probably sustain local area grids, but much of the suburban, rural and long-distance grid would be phased out along with large power plants. Maintenance expenses would decline significantly. Millions of acres of land would become available for other uses. Millions of tons of steel, copper and other metals would be recycled. A few power plants would become museums.

Urban water providers, whether privately owned companies or public agencies, would likely shift from delivering water to managing the watershed. They would oversee groundwater restoration and maintenance of watercourses, as well as external sewage disposal and flood management.

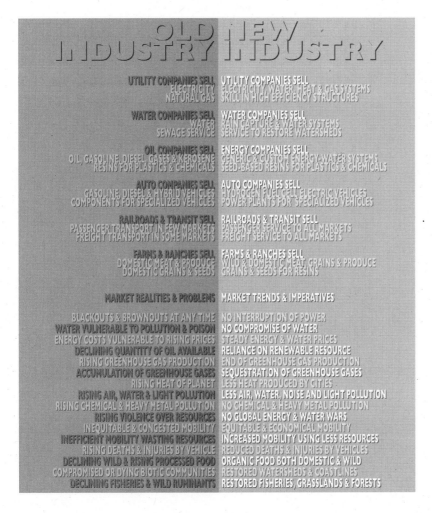

OLD NEW
INDUSTRY INDUSTRY

UTILITY COMPANIES SELL
ELECTRICITY
NATURAL GAS

UTILITY COMPANIES SELL
ELECTRICITY, WATER, HEAT & GAS SYSTEMS
SKILL IN HIGH EFFICIENCY STRUCTURES

WATER COMPANIES SELL
WATER
SEWAGE SERVICE

WATER COMPANIES SELL
RAIN CAPTURE & WATER SYSTEMS
SERVICE TO RESTORE WATERSHEDS

OIL COMPANIES SELL
OIL, GASOLINE, DIESEL, GASES & KEROSENE
RESINS FOR PLASTICS & CHEMICALS

ENERGY COMPANIES SELL
GENERIC & CUSTOM ENERGY-WATER SYSTEMS
SEED-BASED RESINS FOR PLASTICS & CHEMICALS

AUTO COMPANIES SELL
GASOLINE, DIESEL & HYBRID VEHICLES
COMPONENTS FOR SPECIALIZED VEHICLES

AUTO COMPANIES SELL
HYDROGEN FUEL CELL ELECTRIC VEHICLES
POWER PLANTS FOR SPECIALIZED VEHICLES

RAILROADS & TRANSIT SELL
PASSENGER TRANSPORT IN FEW MARKETS
FREIGHT TRANSPORT IN SOME MARKETS

RAILROADS & TRANSIT SELL
PASSENGER SERVICE TO ALL MARKETS
FREIGHT SERVICE TO ALL MARKETS

FARMS & RANCHES SELL
DOMESTIC MEAT & PRODUCE
DOMESTIC GRAINS & SEEDS

FARMS & RANCHES SELL
WILD & DOMESTIC MEAT, GRAINS & PRODUCE
GRAINS & SEEDS FOR RESINS

MARKET REALITIES & PROBLEMS

MARKET TRENDS & IMPERATIVES

BLACKOUTS & BROWNOUTS AT ANY TIME — NO INTERRUPTION OF POWER
WATER VULNERABLE TO POLLUTION & POISON — NO COMPROMISE OF WATER
ENERGY COSTS VULNERABLE TO RISING PRICES — STEADY ENERGY & WATER PRICES
DECLINING QUANTITY OF OIL AVAILABLE — RELIANCE ON RENEWABLE RESOURCE
RISING GREENHOUSE GAS PRODUCTION — END OF GREENHOUSE GAS PRODUCTION
ACCUMULATION OF GREENHOUSE GASES — SEQUESTRATION OF GREENHOUSE GASES
RISING HEAT OF PLANET — LESS HEAT PRODUCED BY CITIES
RISING AIR, WATER & LIGHT POLLUTION — LESS AIR, WATER, NOISE AND LIGHT POLLUTION
RISING CHEMICAL & HEAVY METAL POLLUTION — NO CHEMICAL & HEAVY METAL POLLUTION
RISING VIOLENCE OVER RESOURCES — NO GLOBAL ENERGY & WATER WARS
INEQUITABLE & CONGESTED MOBILITY — EQUITABLE & ECONOMICAL MOBILITY
INEFFICIENT MOBILITY WASTING RESOURCES — INCREASED MOBILITY USING LESS RESOURCES
RISING DEATHS & INJURIES BY VEHICLE — REDUCED DEATHS & INJURIES BY VEHICLES
DECLINING WILD & RISING PROCESSED FOOD — ORGANIC FOOD BOTH DOMESTIC & WILD
COMPROMISED OR DYING BIOTIC COMMUNITIES — RESTORED WATERSHEDS & COASTLINES
DECLINING FISHERIES & WILD RUMINANTS — RESTORED FISHERIES, GRASSLANDS & FORESTS

Major trends in the energy and water businesses, and the automotive and railway industries, as well as organic farming and ecological restoration. All are fundamental civic imperatives. A growing body of citizens, also customers and often influential leaders, recognize that contemporary industrialization is simply not sustainable and that existing industries must make major changes. Trends suggest where industries are headed. In the case of utilities and oil companies they would be selling value-added services, no longer just commodities.

This shift would redefine the role of these organizations to that of overseeing the public commons — the waters.

"Oil" companies are becoming "energy" companies. If oil companies manufactured energy-water systems and related technologies the industry would inevitably prosper. They would produce the means of generating electricity and hydrogen, while retaining ownership of retail outlets — gas stations. What's the value of gas stations that can generate and retail hydrogen, requiring no truck deliveries and no refinery?

From the standpoint of oil companies, their investment in renewable energy represents an investment in changing the nature of the business. Oil companies would gain, in revenue and potential profits, by manufacturing energy-water systems. Instead of selling a commodity only useful to customers with the necessary distribution facilities, or having to invest in the necessary regional bulk storage facilities, pipelines, trucks and gas stations, they would be selling a technology useful to everyone immediately, without need of distribution systems. They would also be building relationships with customers who may later upgrade their systems. These new energy companies would be selling a product where profits are derived from the company's imagination. Instead of playing commodity roulette they'd be playing product poker, where creativity matters.

The utility guy comes to your house and does a survey. Two weeks later the utility gives you and yours a one-week vacation to a nearby resort. Utility guys transform your home. You return and everything seems the same, except it's quieter and the rooms are lit by sunlight even in late afternoon. Next month the bill arrives, it's slightly higher, but it won't be going up and the power won't be going out.

Such a strategy is now possible. Imagine that — the oil and utility and auto companies participating in the development of a new energy-water infrastructure that generates ample energy and water while purifying the world's waters and not only ending the production of greenhouse gases but simultaneously reducing the quantity already in the atmosphere, and by doing this they end black outs and reliance on commodity markets, recycle big power plants and much of the grid, and participate in ecological restoration programs and thus help us all save the world. *That* is profit by all measures.

13

Let There be Light

Wholes are Greater than Parts

THERE IS ONE PROBLEM that has constrained the potential of new infrastructure for decades. It isn't money, nor technical feasibility, nor even the complexity of the problems. It is perception.

In debates over energy, water, transportation or practically anything to do with the commons, the discussion is often heavily influenced by participants who advocate one or another technology. The debate then becomes a nuclear versus solar debate or a reservoir versus desalinization debate. These are valid issues, but structuring the debate on technologies sustains the illusion that our problems can only be solved by another technology, when many issues of infrastructure involve little or no technology.

The root of this focus on tech is the specialized and bureaucratized world of corporations and government. Most corporations are intensely focused on one product line and organize their business in divisions. Most governments are focused on specific objectives and divide any project by subject and time to fit each department. Both corporate and government organizations, analogous to how agribusiness persists in making food fit machines, are forever trying to cram any problem or opportunity into their structure. They shred visionary ideas.

The result of this endlessly finer division of every problem is a fundamental disconnect between how we all see the commons and how those in business or government see it. All people tend to see the world in wholes, we're born that way and must be taught to see and discern the various parts of the world. But we all still see a car, house, road or any other thing in its context. When most people think of a car they speak of it in relation to their lives, not just its price or type or capabilities, but how all those values fit their needs, or don't. Conversely, those in the car business tend to see just the car,

In relation to the commons it can be very difficult for activists, community leaders and citizens, even major corporations, to develop any comprehensive program that involves more than one subject or "division" of information. This is the primary reason why common problems like health care, traffic congestion and land use are not being addressed to any significant degree. While individuals can imagine solutions that may involve energy as well as transportation and water, few institutions are structured to receive such proposals.

or just its price or just the part they are concerned with. They aren't concerned with transportation, just cars, so in any debate they are not going to discuss equitable mobility for all; they're going to tell everyone why cars in general are better, and their car is best.

Government exacerbates the conflict by creating division. In the US and Canada we do not have one retail railway industry handling all kinds of freight and passenger traffic; we have privately financed freight railroads and tiny recreation railways, plus government-financed intercity passenger trains, commuter trains and high-speed trains. Amazingly, government now regards the idea of one company providing all these services as radical. Seventy-five years ago it was called a railroad.

New infrastructure pioneers were and remain a diverse lot: readers of Mother Earth News building a solar home, aerospace engineers captivated with the elegance of photovoltaic cells, utility managers building windmill farms, agronomists who became organic farmers and entrepreneurs who saw wealth in technologies that were smaller, lighter and far more efficient. But they all express one common theme: the infrastructure of the future has to respond to unprecedented global realities. Their views parallel those of their many counterparts who pioneered the practice of holistic health and recognized that human health is inextricably linked to planetary health. Both sought, and still seek, methods of addressing a host of problems *all at the same time*. These idealistic pragmatists saw the potentials as revolutionary not evolutionary. They didn't see a regular building with photovoltaics plugged in, but transformed buildings that use photovoltaics, insulation, new windows and skylights.

There is a new way of viewing our common problems and it's organized around vision. It's more about stories and pictures, less about the brutally confining logic of bulleted points and bottom lines. It's a vision, not a series of platitudes written to fit what someone thought some agency or company might want to hear.

This new view is emerging on the Web. It's visible on countless Websites, where one central vision may be presented with links to every possible relationship. On the Web it's relational, connective, multi-directed and open-ended. Ideas, relationships and money can be quickly organized, linked, guided and decided. Now, for the first time ever, it is possible to design complex projects in virtual space over all continents involving any number of people without the arbitrary structure imposed by government or corporate decision-making. Now we can all decide what technology we want in our lives, rather than be told how we must change our lives to fit some technology.

Ending War by Vision

Vision is the source. Arising by prescience a vision can be a personal statement of what one wants to be and wants to create. Such personal visions can align over time to become a common vision, or a vision of the commons.

This seemingly simple relationship between vision and reality is at the core of what it is to be human. We all do what the animals do, with toilet paper. But we also do what no animals do, with writing paper. We envision our tomorrow. We cannot change the physical world instantly, but we can change our vision in a millisecond. Ideas change our world.

In many parts of the world, in both developed and undeveloped regions, millions of people face a daily struggle to survive, while millions more live with the uncertainties of not knowing whether their job, their whole career, or their whole community will be decimated by some technological, economic or political shift over which they have no control. Increasingly this uncertainty is extending to all levels of society, not just those with the least monetary wealth.

The focus of the world's media on specific events, notably the war in Iraq, conveys uncertainty to all, and in the process seemingly blinds us all to the certainties and realities of the region surrounding Iraq. These certainties involve a movement, in cities, towns and regions towards new infrastructure, specifically railways and solar-energy. These plans were not imposed by other nations but emerged spontaneously from local councils and business and civic leaders; they are not a recent program, but have been evolving for over a decade.

The Middle East, by the broadest definition, extends from the Nile to the Indus, from the Hindu Kush to the Arabian Sea and from the Caspian Sea to the Gulf of Oman. This area includes major oil-producing countries and most of the world's opium, holy sites recognized by three major religions and innumerable archeological sites. It includes 15 nations defined by boundaries imposed by former colonial powers. It is largely a sea of sand dotted with islands of mountainous desolation where villages cling to geological edges. It includes places called "Empty Quarters."

The world takes from the Middle East as if it were all an empty quarter. A world of people know little of the region and have no reason to care, except for how any conflict in the Middle East might disrupt the flow of oil.

We are all faced with global problems unprecedented in scope. The Middle East is especially vulnerable given its minimal water supplies and declining oil. Simultaneously the rest of the world, until weaned from oil, is vulnerable to further uncertainty as oil supplies diminish. Stability in the Middle East is in everyone's best interest.

The nations and cultures of the Middle East do not share a common vision of infrastructure. Highways and airlines serve major cities, but not many smaller towns and villages. Cars are viable in cities, but long-distance desert highways are minimal and often crowded. Trucks move nearly all the freight, but trucks are costly. Water systems are often minimal or just old and leaky, and in many areas are major sources of contention. Energy systems are diverse, with a small but rapidly growing renewable energy industry centered in Israel. Overall, the region's infrastructure ranges from kerosene lamps in mountain villages accessible by trails used by local traders astride donkeys to new urban complexes lit by natural gas and accessible by freeways where global traders commute to work at electronic consoles in places like Dubai.

Any comprehensive vision of new infrastructure today begins with transportation: the means of propulsion or the entire system. Transportation is the key technology because it's the most widely understood, most widely valued and most widely needed and because it has historically been the vital investment that justifies and triggers investment in all other infrastructure. Inevitably a railway, highway or airport requires passenger stations, gas stations and terminals. Each of these facilities inevitably becomes a node of activity, and all the technology used to support the buildings and services tends to be replicated in the commercial and private structures around the station, off-ramp or terminal structure. Thus the system seeds infrastructure at every node.

Beginning in the 1990s, with government initiatives in several countries, the greater Middle East began to define what may, in the next decade, result in a common vision of transportation. There was no particular plan, rather just a series of isolated projects confined to particular countries. Planners in Egypt, Israel, Iran and Dubai (one of seven United Arab Emirates) began recognizing that growing and more centralized populations would require major transportation investments, and that the region's population is spread over vast distances of desert, necessitating long drives across empty quarters where sandstorms clean the paint off one's car. Railways are receiving much more attention.

By 2000 several railway projects were being completed in the region, in Israel and Egypt, while plans were being drawn for more routes. The Saudis

Developing infrastructure may be the most critical step in building a peaceful world. While the immediate benefits are most visible to users of an infrastructure, the far more profound implications are barely noted. Railways, more than highways, initiate development of communications, energy and water systems in the places they serve. These services make possible unprecedented business and cultural relationships, and this means trust between former strangers grows. Sometimes love blossoms and genes meet. War becomes unthinkable because the enemy includes your in-laws, or your customers, or your lenders.

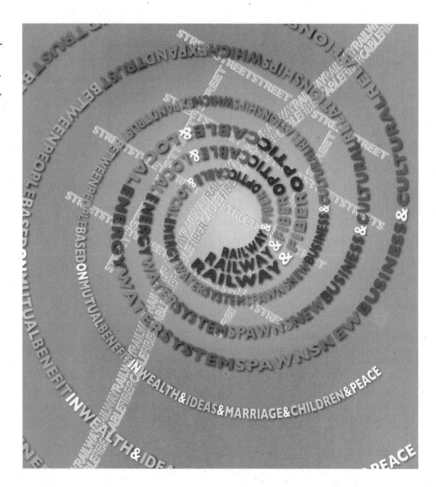

are building a 600-mile expansion of their modest existing railway, with a "railbridge" across the country from the Persian Gulf to the Red Sea, and new lines to Medina and Mecca to provide transport for hajj pilgrims. Meanwhile Dubai is building a 43-mile metro railway; Israel is rapidly expanding their railways, including a new line to Jerusalem; and China is angling to build a railway west via a route that would link with the railways of Iraq and Turkey. And, before Saddam Hussein was toppled, Iraqi and Iranian planners were planning a high-speed rail link between Tehran and Baghdad, and a line north to Russia.

China is interested in a railway version of the ancient Silk Road, the southern counterpart to Russia's Trans-Siberian Railway. There has also been talk in railway industry circles of connecting the railways of the Middle East with those of India, Southeast Asia and China via new track in Afghanistan and Myanmar (Burma). Either or both routes would be formidable projects, but one or both are perhaps inevitable, simply because they represent access to European markets via Istanbul, and that means major savings in time, resources and money compared to sea routes. A US-style railway, with double-stacked container trains, would radically reduce ocean shipping via the Suez Canal. No doubt millions of passengers would ride segments of the route, while an adventuresome few would ride the whole distance. Once a day there could be a true "Orient Express," a sleeper train leaving London for Beijing, via Paris, Venice, Istanbul, Ankara, Baghdad, Tehran and on across the sub-continent — a truly Grand Tour.

All the projects now in planning or construction throughout the region represent the skeleton of a regional system. New east-west mainlines could link the railways of Pakistan and India with the railways of Turkey and Europe as well as Egypt and south into the Sudan, potentially all the way to Capetown. North-south mainlines would link the region to the north via Kabul to Tashkent and Tehran to Tbilisi; and the cities of the region from Ankara via Beirut to Cairo. There could be branch lines to smaller towns, scenic areas, sacred and historic sites. Freight service would reach 70 miles per hour on many routes; passenger service would range from 120 to 150 miles per hour on major intercity routes.

A modern railway by definition requires a parallel communications line, an energy system, a water system and a series of stations. In contemporary terms this usually translates to a fiber-optic communications line paralleling the railway; an electrical system of overhead wires to power trains, using electricity bought off the grid or generated by a dedicated power plant (alternately a diesel fuel-distribution system for locomotives); and a water system for employees and passengers, with sufficient quantities for washing equipment, dining service and toilets — mobile and stationary.

What if the entire region linked all existing and planned railways and defined the system around state-of-the-art energy, water and communications technology? Stations would become the nexus for local transit, taxi, bus and truck services, as well as high-speed voice and data transmission systems. Stations and trains would use new energy-water systems and biological sewage disposal, thus providing the basic infrastructure to support local purchase of hydrogen-electric cars and trucks, in-house power systems and neighborhood-scaled sewage facilities that produce useful fertilizer. A truly regional system, reaching all cities and major towns, as well as major facilities like airports and public attractions, would involve around 2,000 stations scattered over 15 countries. Every station would demonstrate the latest technologies to the countless citizens and visitors who ride the trains.

How would the development of such a vast system affect the region? In economic terms freight and passenger transportation would be uncommonly inexpensive, and energy prices as a portion of total transport and living costs would be a fixed finance charge rather than a local or global commodity price. Industries could be extremely competitive on global markets. In cultural terms it could dramatically increase regional travel, crossing many cultural and political borders, and possibly even reduce regional tensions. Tourism would no doubt increase since the railway would reduce the need of long boring drives, and the negative impacts of tourism on cities, towns and desert ecosystems.

For over 20 years the Saudis, and others in the region, have been active in solar-energy development because some groups believe it's imperative the region convert to renewable power before their oil runs out. While the debate rages over how much oil is left one fact is unavoidable: it will run out. Building a solar-energy and water infrastructure in concert with a railway system would accelerate the transition away from oil, and potentially reduce or end many of the conflicts that arise over oil and water.

One might think such a major program, involving railways and new technology over a vast region where people speak at least one of four major languages and worship one of three major religions, and where an ongoing war remains the region's focus, is borderline impossible. But history and the reality of the Web suggests otherwise. In the history of infrastructure there are ample cases of whole transport systems being built before, during and after a war, much like experienced computer users "work around" a flaw in a software program. Moreover, the Web makes it possible to bypass many government-to-government problems by directly linking all involved at the local and global levels. Thousands of people could interact across all boundaries in several languages. They could design their future — literally.

What if this future is already developing? While battles might rage in small towns and urban alleys in Afghanistan, Iraq or the West Bank, it would seem a new infrastructure program is slowly coming to the surface via a few Websites. What if now disparate plans were assembled into a regional system? Could the vision represented by rail, communications, energy and water infrastructure — and the visions it spawns among countless civic and business leaders — evolve to become a widely shared vision of the region's future? Could such a comprehensive vision represent the best means to move beyond the conflicts that have plagued the region for decades?

An Economy of Vision

Regions like the Middle East are wrestling not just with simmering unrest, but with profound questions about their future. Many people in Africa, Central and South America, India and Asia also see a future of endless uncertainty. Millions of rural people are not moving to big cities just to find a job, but to find a future. Cities have grown to unprecedented size largely because the uncertainty in the country is greater than the uncertainty in the city.

Aid agencies have long assumed that the best antidote to urban migration is rural infrastructure. Unfortunately, all too often the infrastructure programs have been oriented to massive dams or other big projects that were more about the donor nation's need to make a statement or a contractor's desire to make big bucks, than the community's real water and power needs.

New infrastructure technologies are inherently decentralized and have already made major inroads in less developed regions, where cell phones, photovoltaics and computers are seen as doorways to the future. Unlike previous programs, where one massive infrastructure project was an unavoidable first step, the new technologies can be developed in large or small increments wherever needed.

New infrastructure technology offers the means to provide the basics to the humblest of communities. This means lights, television, computers and

An energy-water system can be developed for all 6.5 billion citizens. All systems would cost $13 trillion ($3,000 per capita), equal to less than three percent of the world's $65 trillion gross world product (GWP) as expressed in debt service. This would represent energy and water for individuals; commercial and transport systems would be developed in parallel. It is assumed 80 percent of water used is recycled, yielding demand for new rain/well water equal to a quarter of existing water consumption. Assuming a low seven inches of rain annually one person would require 400 square feet of roof to capture water, with photovoltaics covering a quarter of that area. The roof area needed for all people would be equal to a square 300 miles to a side, or about two-tenths of one percent of Earth's land area.

ENERGY-WATER SYSTEM FOR ALL PEOPLE & HOMES OF THE WORLD

TOTAL POPULATION	6,500,000,000 ('07)
GROSS WORLD PRODUCT	$65,000,000,000,000 (")
GROSS WORLD PRODUCT P/CAPITA	$10,000 (")
GALLONS HOUSEHOLD WATER P/YEAR	44,600,000,000,000 (")
GALLONS PER DAY PER CAPITA	18.8 (")
TOTAL CONVERSION CAPITAL	$13,875,000,000,000
AVG PRICE PER KILOWATT HOUR	$.17
EXAMPLE PRECIPITATION	7 INCHES
GALLONS HOUSEHOLD WATER NEW	11,500,000,000,000
GALLONS PER DAY PER CAPITA NEW	24.7
SQ FT ROOF TO CATCH WATER P/CAPITA	400
SQ MILES OF ROOF TO CAPTURE WATER	95,000
DIMENSION OF SQUARE IN MILES	300 x 300
SQ FT ROOF TO ELECTRICITY P/CAPITA	90
SQ MILES OF ROOF TO ELECTRICITY	20,000
DIMENSION OF SQUARE IN MILES	144 x 144

YEAR	1	2	3	4	5	6	7	8	9	10	11	12	13	14	15
CUSTOMERS P/YEAR	3M	7	33	65	156	260	520	580	650	715	715	780	780	715	520

Following a one-time design program of a few years
Number of homes converted to 100% solar-hydrogen energy-water system in year

CUSTOMERS TOTAL	3M	10	42	107	263	523	1.04B	1.6	2.3	2.9	3.7	4.4	5.3	5.9	6.5

Number of people served by systems developed, including children

JOBS/INSTALLATION	0.7M	2	7	14	34	58	116	131	145	160	160	174	174	160	116

Millions of people employed in installation of systems in homes

CAPITAL PER CAPITA	$3.0K	$2.9	$2.9	$2.8	$2.5	$2.3	$2.2	$2.1	$2.1	$2.1	$2.0	$2.0	$2.0	$2.0	$2.0

Price of buying & installing energy-water systems, declining by innovation & economies of manufacturing

TOTAL CAPITAL YEAR	$9B	$19	$94	$181	$396	$594	$1.1T	$1.2	$1.4	$1.5	$1.5	$1.6	$1.6	$1.5	$1.0

Capital invested by utility companies in buying new systems, per year, in billions & trillions

AVG BILL PER CAPITA	$22	$22	$21	$20	$19	$17	$16	$16	$16	$16	$15	$15	$15	$15	$15

Average monthly bill would remain the same as a fixed finance charge, no matter how much power used

AVG COST PER KWH	$.22	$.22	$.21	$.20	$.19	$.17	$.16	$.16	$.16	$.16	$.15	$.15	$.15	$.15	$.15

Price of electricity only, no value assumed for pure water and heat

TOTAL REVENUE YR	$874M	$2.6B	$11	$27	$62	$116	$217	$330	$454	$589	$723	$869	$1.0T	$1.1	$1.2

Revenue to utilities for financing and maintaining the systems, from millions to billions to a trillion

PRCT GWP P/CAPITA	2.7%	2.7	2.6	2.5	2.3	2.1	1.9	1.9	1.9	1.9	1.9	1.9	1.9	1.9	1.9

The cost of this system would be insignificant in proportion to gross world product refrigeration, although not necessarily in that order. Light, TV and computers facilitate culture and business. Refrigeration and energy for cooking mean higher quality nutrition and the ability to keep medicines. In low income communities refrigeration and pure water are not merely symbols of a better future, but of a future where children are less likely to die by dysentery or other preventable diseases.

Financing new infrastructure for a village, city or whole region of modest income is generally viewed as the single biggest barrier to change. It's assumed the only possible option is a grant, usually from a rich nation or nations, sometimes with corporate and/or private foundation contributions. Conventional aid programs emphasize grants only, which can often undermine self-reliance and cause recipients to resent their benefactors.

People want the opportunity to create their own vision and giving that up is to lose hope and liberty at the same time. They want capital. Regardless of arguments about capitalism common in progressive political circles, those in less developed circumstances find capitalism works. But it must be recognized that capitalism is expressed in a diversity of ways, with the cowboy brand as practiced in the US being widely disdained.

A new bank was started in Bangladesh in 1979 and it has changed the world. It is called the Grameen Bank, meaning "Bank of Villages," and it now has 2,185 branches in 70,000 villages in Bangladesh. The total number of borrowers is about 6.5 million; 96 percent are women. The loan repayment rate is over 98 percent, significantly better than most banks'. The loans may range from $5 (US) to $100 or more. Since the bank's founding they have loaned $4 billion, and in most cases borrowers were able to dramatically improve their situation. The Grameen Bank has grown to include communications, energy, education and weaving subsidiaries, all designed to assist the poor. The bank is 94 percent owned by the borrowers.

The Grameen Bank triggered a microbank movement based on the simple premise that people were not a bad risk just because they had little money. There are now similar microbanks in 40 countries. Notably, two products are often at the top of the shopping list: cell phones and photovoltaic cells. The Grameen Bank has sold over 140,000 phones, many being purchased by women who become the village phone company. Photovoltaics are sold to power the phones. Talking by light.

New infrastructure is made far more feasible by the existence of microbanks. Once a modicum of basic infrastructure exists the stage is set for the growth of businesses and community economies in rural and urban areas. Plus more banks.

New infrastructure could set the stage for "eco-development." Sustainable businesses can be built around existing ecological resources. In coastal regions this might translate to restoration of wetlands, marshes and estuaries, and consequently river and ocean fisheries, which can result in businesses built around fishing and natural fibers. In other regions new infrastructure might facilitate game ranching, low-impact logging or hunting for new chemical compounds in the life of jungles.

While new energy-water systems are likely to remain a significant investment, the ratio of cost-to-benefit would be far better than existing technologies. Plus the community controls the technology, not a remote agency.

No one can know the latent vision present in the peoples of Khandahar, Havana or the Bronx — except them. It's increasingly possible for everyone, rich or poor, to have access to the tools to express his vision. In developing infrastructure we develop the commons, and only what's truly equitable is truly of the commons. In that commons we all benefit if people can design their own futures, create their own visions and build an economy on light.

14

Crystals, Light and Living Waters

Razor's Edge of History

WE LIVE IN AN AGE of serial doomsdays. It's as if millions of people became addicted to the Cold War and ever since it ended they've been scrambling to invent a new enemy. Some choose terrorists, some blame corporations, others blame the secular society and still others seem to harbor a profound guilt and therefore blame humanity — equal opportunity blame. This seemingly endless focus on seeking blame is fed by a mass media industry that is so desperate to capture eyeballs it is willing to foment any conflict to attract attention. It does so in industrialized societies that seem to have lost comprehension of their own capabilities, ignored their own native wisdom as expressed by countless citizens, and instead indulged in bleak visions of naked fear as exemplified by movies like *Blade Runner* and its countless imitators.

If we see beyond the realities of climatic and ecological change, ignore the childish fantasies of technocratic futures, and look to the trends and technologies that have emerged in the past few decades, we can discern a far more optimistic vision shared by millions of people. Their stories will not be on TV, as they've been ignored by mass media, but they are present in magazines, Websites, companies, nonprofits and, increasingly, whole industries.

These people share a view of life that is far more inclusive than any conventional cultural or religious beliefs. Their visions are about the air, water and sun — what we all share.

"Religio" is Latin for an obligation of reverence, and "religare" means to bind. It could be said religion is the tie that binds us to an obligation of reverence. In contemporary life religion is usually discussed in relation to worship of an abstract being, rather than reverence for all life, with its natural cycles and rhythms. But this new view expresses a religious reverence for the planet as if the being were not separate from, but a part of.

This larger religious view crosses secular lines and blurs the boundaries of all organized religions. It is expressed in a variety of ways not easily characterized. However, certain elementary ideas are common: notably the concept of reciprocity expressed by the Golden Rule — do unto others as you would have them do unto you. Most of us most of the time live by reciprocity. It is the essential ethic of civilization.

Reciprocity is an imperative in sustaining our ecological balance. We know taking more from fields or seas diminishes our food supply, but we don't all recognize how we are now destroying our *means* of producing food. We are destroying not just the diversity of biotic communities but also the fertility of the soil that supports them. In effect we've been withdrawing money from the bank in excess of our deposits. We've been taking far too much and giving back far too little.

Sustainability can only be achieved if we put lots of money back in the natural bank and live on the interest. Restoring the Grand Banks or the Great Plains is an investment we know will return dividends in sustainable yields of several species, but no one knows how much. Restoration is equivalent to investing in earth, air, fire and water purely on faith that the profit will exceed the investment. Our survival would be the profit.

It is a fundamental reality of all natural systems that in each is all and in all is each. This doesn't just mean the fish you ate is inside you and the fish he ate inside him; it also means the sea of air, water and minerals that flows through you and all the fish right back to the headwaters where the littlest fish was born five months ago in the shade beneath a pine tree. It also means that if rising temperatures kill the tree there will be no shade and the little fish will never be born. It is all one contiguous pattern of water and life spawned by the sun, driven by thirst and a will to live that can only be traced to inexplicable and incomprehensible origins.

The problems we face are pervasive and demand uncommon participation by all. The sun's energy is decentralized, life is decentralized, the problems are decentralized, cities are decentralized and we are decentralized, therefore our response to climate change and the peaking of oil must be decentralized. It must arise in all forms in all places simultaneously. It already is.

Life is decentralized. The vast majority of electrical demand involves small structures requiring a few to several hundred kilowatt-capacity service, while a relative few might require a megawatt or more. Water usage is similar, with millions of small customers scattered over a vast area and most using a few hundred gallons a day, not millions. Rain and light are also decentralized. Yet despite these realities it is widely assumed we must continue to rely on large centralized facilities.

Addressing energy, water, transport and land use issues demands small quantities of people-power all over the place rather than large quantities of government-power from any one place. And just as small quantities of power can be generated by light all over the place so can modest quantities of people-power be generated by faith all over the place. But our faith is limited. On the cosmic faith-o-meter we all have just so much.

Sunday could become Sun Day. One day for all. Millions of people who share a love of the wild often go to parks on Sunday; some might have gone to church in the morning while others may view the wilderness as church. In any case Sunday has been blessed. Sun Day for the commons.

Communities all over the world are taking action to address concerns of the commons. Whether people would describe themselves as secular or religious they are all working for a greater good. They might be engaging in complex research on their own time, or restoring a creek as a class project, or developing a new farm on a shoestring budget, but all sustain the effort, as many have for decades, by their faith in humanity.

Their faith is built upon recent successes and historic precedents. The successes embrace major trends involving large corporations, such as the growth of renewable energy, big projects like the restoration of Buffalo Bayou in Houston or even the little piece of Berkeley's Strawberry Creek liberated from a pipe and now supporting trout.

From our vantage at the beginning of the 21st Century it would seem we are faced with essentially two pathways. One is a well-worn path while the other is a new path as yet vague.

The first path is often defined by established institutions based on the premise of control. Nuclear power is on this path, as is genetic engineering and a host of strategies and technologies rooted in the belief we can and should control nature. This path has resulted in an environment permeated with heavy metals, inorganic chemicals and radioactivity. This path is built on facts and driven by breathtaking hubris in the assumption that human beings can know everything, that uncertainties can be managed and that our collective future is little more than a continuation of past history.

The second path is defined by the premise that we are not in control. Organic food is on this path, as are industrial ecology, solar-energy, green architecture and ecological restoration. This path has resulted in air and water becoming cleaner due to pollution controls, the development of a panoply of new energy technologies, adoption of recycling as a standard practice and a renaissance in a global railway. This path is built on facts tempered by faith, and is driven by a profound sense of vision, aligned with a recognition that our collective future is influenced by all people, yet determined by the grace of the divine.

One path sustains an existing concept of industrialization that is only accessible to some people, while requiring centralized control, as exemplified by nuclear power. The other is a path leading to a new realm of industrialization accessible to all people, with no need of centralized oversight, as exemplified by photovoltaic cells. This path is already visible in the computer-driven transformation of printing, motion picture, music and other media industries, as well as manufacturing and materials technologies. This path gives individuals and companies the ability to engage in a wide range of businesses using far less initial capital — the price of entry.

We are on the razor's edge of history between a path of the past with all its knowns and a path of the future with its myriad unknowns. The path we choose in the next decade will have ramifications extending far into the future and with implications for virtually all life on the planet. Our choice will be wise only if we are armed with facts and guided by faith. What would the children of today want us to do?

Meanwhile, Back in the Future

Intelligence is usually defined as only individual — one brain at a time. Yet it would seem there is a group intelligence where many people tend to "know" what others are thinking or doing. Group intelligence could extend to group prescience, where many people share a vision rooted in a future that appears to have no bearing on contemporary realities.

Concern over the impacts of technology in the sixties was motivated by a profound sense something was wrong with a society that believed it could foul its nest without consequence. This concern caused thousands of individuals and small groups to drop out and do their own thing, whether it was solar architecture, organic farming, renewable energy or dozens of other technologies and initiatives. They all shared a recognition that what was happening could not be sustained.

Their views were widely derided as childish fantasies by a mainstream culture that thought everything was okay. When nuclear power was going to give us electricity too cheap to meter, as advocates promised in the 1950s, who would have believed it never would, and a metallic crystal would lead to homes and cars powered by light? When it seemed everyone was moving to the suburbs who could have imagined thousands of small towns reinhabited by a new generation? When dams were shining examples of technological wisdom who could have conceived a dam being removed in favor of fish? When the jet set was jetting who would have imagined trains making a comeback?

They certainly didn't learn these things in high school. Where did they get these ideas? Who was that masked man?

Why did a large portion of one generation go off on another path and do so knowing they were going against the grain of modern society? Why were they compelled to spend years pursuing ideas that seemed crazy to everyone else, even though they achieved social ostracization before they achieved an income?

Prescience means to know something will happen before it happens. It is not a matter of mere educated guesswork because the work of countless visionaries is based not on existing technologies, but seems to come out of the blue. Jules Verne was a writer, yet the US Navy credits him for conceiving the periscope as described in *20,000 Leagues Under the Sea,* a novel he wrote several decades before the first practical submarine was even contemplated. Countless visionaries have spoken of a vision coming "through them."

Given the likelihood of failure due to a myriad of calamities that can befall us all, why would anyone undertake a life premised on a vision unimaginable to anyone else? What other force but prescience could explain why millions of people would opt for a future of uncertainty based on a fantasy of harmony?

Millions of Americans and millions of citizens in other nations have manifested their visions of life over the last few decades. Mass media have packaged these trends as the "self-actualization" movement, the "new age," or related movements for "voluntary simplicity." Reality has outgrown the monikers. The voluntary simplicity self-actualization of the new age is now a way of being that transcends gender, age and income. It's do-it-yourselfers of the fifties morphing into do-your-own-thing folks of the 70s morphing into just-do-it people of the 90s morphing into just be-it people of the 21st Century. Whatever your particular "it" is, you go for it.

Humanity hasn't lost its vision: it just doesn't know it has one. This "greening" of America, this new age paradigm shift, and this wholly new amalgam of cutting-edge high-tech with ancient low-tech and subtle no-tech all represent facets of the same diamond. Millions of prescient beings saw it decades ago, and millions more have joined the show, and it's now a vast and complex culture engaged in a myriad of activities. Their visions are not orthodox by the standards of any particular group. They are *their* visions.

Yet they share a vision of a sustainable society actualizing far greater human potential on far less of all resources, with far less violence and far more respect for the inviolate nature of water and air and the plants and animals by whose grace we survive. This vision is all of a piece because it is bound by common antecedents to all religions and all peoples. This vision is beyond all precedent in its recognition that we, as a species, have reached the end in the creation of weapons of ultimate destruction. *We* are the weapons of mass destruction.

Millions of people have seen the horror of Armageddon as clearly as they see their own eyes in a mirror. Millions more, living the despair of slums in any of the world's mega-cities, live Armageddon daily. Yet despite these knowns some millions of us saw something else, some path to transform the situation, that was utterly unknown. A new path that might just change the world.

Our exploration of that path could only be defined by prescience because there are no maps. Calling forth this knowledge — call it imagination or call it divine prescience — is an act of reverence. By this vision we can see what's never been seen before, and we can all do it. The vision can happen in a second or a lifetime; it can be as simple as defining a plan to change a kitchen or as complex as designing a program to transform a city. Curiously it's commonly said "you can't change the world" when in fact we do just that and we do it almost compulsively. It starts with a vision.

Seeing a future, horrific or beatific, can be a moment of resignation or creation — it's a choice. Resignation to the demise of all civilization is a conscious act of creative effort. The mind must build a story, replete with all the justifications, hideous examples and logical constructs, to justify the view that humanity is over. It's a conscious act of self-fulfilling prophecy. Choosing to believe civilization and all forms of life will survive and prosper is also a conscious act of creative effort. The mind must also build a story, of logic and precedent and ideal. Both views are statements of faith. Which prophecy would you like to fulfill?

Right now the issue is faith, not time or money. If we have the faith we can make the time and find the money. Money is what you get after the check clears if you have enough faith to show up.

Diamond Light

Once upon a time, not all that long ago, virtually all the waters were pure, all the seas were full of fish, all the forests were rich with diversity and all the people survived on a land teeming with life. We were all organic.

Now we are 6.4 billion folks living all over the planet in vast urban complexes that consume staggering quantities of energy and produce prodigious heat. Now we're faced with shrinking forests, declining fisheries, drying land, a question of energy sources and a climate-change endgame with a methane trigger — unless the gun freezes.

Yet now we are also a population that increasingly knows it is one. Just as a child gains consciousness of self as he or she grows, so also has some large portion of the world's population gained a consciousness of humanity enfolded in the universe. These people know we are not apart from nature, but inextricably woven into the natural cycles of life and the ebb and flow of water, air, fire and earth. They know a river runs through us and so does electricity. We are electric water.

Paradoxically, our forays into space, intended to explore the world out there, gave us a window on the world in here. American astronauts offered us snapshots of Earth, telling us in photographic truth that we were indeed on a blue ball in space, but they also spoke in reverential terms of the sheer beauty of the place. Russian cosmonauts, after months inside the Mir space station, became testy and had difficulties working with each other. One of their science experiments required they grow shoebox-sized plots of grasses.

Taking care of their tiny fields was the only activity that kept them sane. It kept them connected to home.

Animals are our link to the life of the Earth beyond our skins. For the first time people are talking to animals — chimpanzees and gorillas in captivity — using sign language. People have also developed relationships with wild animals, most notably several species of cetaceans — dolphins and orcas. Our ancestors had close bonds with many wild animals. Domestic cats and dogs, as well as cows, sheep or pigs are living proof.

Environmental problems are often discussed as if they were happening to someone else somewhere else. Global warming, or anything concerning the atmosphere, is by definition a concern of all humanity. No matter what our beliefs or position in society or place on the planet we are all living within the thin bubble known as atmosphere. If we could drive vertically "outer" space is a mere hour's drive away.

In the fall of 2005, just off the coast of San Francisco, a migrating gray whale was swimming south. She was spotted by divers in a small boat as she struggled to breathe. One fluke had become tangled in a fishing net and it was dragging her down. The divers grabbed their knives and dove in. After several minutes cutting the ropes the net fell away and the whale surged forward, suddenly free. She began to continue her course, then turned around and came up to the divers and touched them. The divers could not know, but they felt she had thanked them. What other explanation could there be?

Such experiences, often shared with billions of television viewers all over the world on countless wildlife shows, reinforce our ancient linkages to all species. Perhaps in this profound conversation with animals we are seeking the guidance of spirits buried deep in our collective memory. We cannot know how animals perceive reality, but we do know they play; they can be happy, sad or mad, and they can evoke by their eyes a mystery we cannot fathom. By their behavior we can know the state of the planet and by their presence we can be sure its life is sustaining itself.

Hawks nest on Fifth Avenue high-rises in New York City. Golden eagles have returned to San Francisco Bay. They now use the warm updrafts off apartment building air conditioners to climb higher and higher by spiraling up. Sea lions have taken over a park in Monterey —they even use the benches. These are signs the Earth can heal.

Animals evoke the reality of ecological memory and the value of our own history. We can know by recent history that the seeds of forest and grassland, reefs and kelp beds, are all still here and with a modicum of assistance on our part these ecosystems can return to abundant life. From geological records spanning millennia we can know a great deal about climate, ecology and our existence. In this data and in our collective memories and stories we hold the evidence of what was. Our only guide to what the planet could be like lies in our memories of what was. Our memory is our lifeline to our future.

We cannot return the planet to its ecological state of centuries ago, but we can restore many of its myriad ecosystems to a far, far more productive and healthful state, where our lives, our cities and our economies are not contrary to the natural world, but are bound to its cycles and woven into its wildness. This can only occur if we end our allegiance to unsustainable physical growth and remove the barriers to unimaginable natural growth. The barriers are not merely physical structures, as exemplified by dams, but also belief systems where it is presumed that people always know better than nature, that fish farms are more efficient than rivers, feedlots more efficient than grasslands and the controlled is more efficient than the wild.

We are faced with a headlong rush into the unknown, of technologies already used and technologies yet to be used, of atmospheric and ecological changes of staggering magnitude. It would seem our only bottom line is wondering how an idea or technology that is based on our memory of what was and our imagination of what could be can enhance the sustainability of civilization and health of the Earth.

Our visions of change will only succeed if we increasingly recognize all the facets of our actions, both what we've done and what we envision doing. We must consider the people, animals and plants; the past and the future; and all the myriad relationships between earth, air, fire and water. We must recognize how nature seeks a balance, and will seek that equilibrium with or without us; thus our only choice is to seek that same balance by our actions. Everything is linked and everything matters and there is a unity in it all.

If there is one object that symbolizes the historical moment it is a crystal. Crystalline structures made of silicon can transform sunlight into electricity. Crystal lasers now propel our voices and pictures at lightspeed through fibers of glass. Crystal lasers can accelerate data in computers by using photons instead of electrons.

Quartz is the source of silicon. Remove the oxygen from quartz and you get pure silicon. Melt the silicon in a crucible. The resulting silicon ingot will naturally organize itself in a diamond pattern. If you slice the silicon into wafers, add specific impurities, and attach some wires, you then have a semiconductor. If it's a logic semiconductor you can print millions of tiny circuits on one wafer, then cut them into little chips for use in computers. If it's a photovoltaic semiconductor you can place it in the sun and it will generate a watt or two.

Glass begins as sand, preferably white sand with a high quartz content, plus soda ash found in dry lake beds. These two materials are melted at about 2,500 degrees F until they become a clear liquid. Poured on a pool of molten metal and cooled, it becomes plate glass, poured into a steel mold it becomes a bottle, or drawn into hair-thin strands it becomes fiber-optic cables. Glass has long evoked the notion of crystal purity and clarity ever since the first person melted some sand and saw right through it.

That which we describe and discuss via our computer chips and fiber-optic lines is often structured like a crystal. Countless drawings of ecological systems and software systems and organizational systems are defined by relationships between different ideas, species or people drawn in forms resembling crystals, often within circles to convey relationships between all things within a greater whole. Crystalline structures visible in the art of many religions are expressions of the divine unity of the universe.

Silicon Valley was built on crystals. Lying on a beach on the California coast and staring at the sand one can see the tiny grains of nearly transparent white quartz. They sparkle in the sun. The quartz comes from a range of mountains to the east, where it was eroded from blocks of granite streaked with veins of white quartz often mixed with gold and silver. In those tiny grains is the raw material capable of transforming light into electricity.

We are on the threshold of an age of light. The technology exists to power our world largely by crystalline structures of incomprehensible elegance and unimaginable size. It will be a world where all life prospers by light and rain. Like us, again.

In 1967 a hippie was selling quartz crystals. His sales pitch: "The future's crystals man."

Resources

The following Websites represent some of the sites and organizations referenced in writing *Electric Water*. These sites offer the broadest range of information and links to other sites. Searches of the following categories will generate far more results than listed here.

Searchable Topics

- ecological restoration
- global warming
- solar-energy
- solar thermal power
- fuel-cells
- proton exchange membranes
- hydrogen power systems
- automotive battery power
- freight railroads

- organic farming
- peak oil
- solar heating
- photovoltaic cells
- electrolyzer
- metal hydrides
- lithium-ion batteries
- electric cars
- passenger train

Energy-Related Sites

US Department of Energy, Energy Information Administration
eia.doe.gov

Annual Energy Review, US Department of Energy, Energy Information Administration,
eia.doe.gov/emeu/aer/pdf/aer.pdf

Household energy report
www.eia.doe.gov/emeu/reps/enduse/er01_us.html

If you want to know where energy comes from and where it goes this is the place to start.

Lighting Research Center, Rensselaer Polytechnic Institute, Troy, NY
First-rate site with access to information and career opportunities.
lrc.rpi.edu

Energy Conversion Devices
Stanford and Iris Ovshinsky are the founders of Energy Conversion Devices. They invent industries rather than products.
ovonic.com

Sanyo
Sanyo is a Japanese company of stunning achievements See their "Amorton" photo-voltaics site for products both rigid and flexible.
us.sanyo.com/home
semiconductor-sanyo.com/amorton/index.asp

Kyocera Solar Inc.
One of the largest producers of photovoltaics in the world.
kyocerasolar.com

Open Energy Corporation
A new company that is developing a non-imaging concentrator designed to gener-ate high-temperature steam.
openenergycorp.com

Consumer Energy Center, California Energy Commission
This site provides considerable value with its explanations of technology and its im-ages and links.
consumerenergycenter.org

Pacific Gas & Electric Company, California
This utility has long been involved in photovoltaic and related solar programs.
pge.com/solar

Solar Buzz
Latest prices, trends and lists of products. Your first stop for any overview of the solar field.
solarbuzz.com

Photovoltaic Energy Systems
This is the site of Paul Maycock, *the* expert on the photovoltaic industry and trends.
pvenergy.com

Ceres: Investors and Environmentalists for Sustainable Prosperity
A coalition of major corporations engaged in projects to increase sustainability.
ceres.org

SolFocus Inc.
Innovative company developing solar collectors using mirrors to focus sunlight on photovoltaics.
solfocus.com
parc.xerox.com/research/projects/cleantech/cpv.html

Ballard Power Systems, Inc.
Leader in producing commercial fuel-cells as back-up power and for cars.
ballard.com
ballard.com/resources/animations/animations/FuelCellShort/main_
content.html

Schatz Energy Research Center, Humboldt State University
Humboldt State's College of Natural Resources and Sciences is a major center for environmentally-related science. One building runs on a photovoltaic-hydrogen-fuel-cell system.
humboldt.edu/~serc

Lambda Energia
A photovoltaic company engaged in development and manufacture of thin-film photovoltaic cells and related products.
lambdaenergia.com

US Department of Energy, Hydrogen page
A good entry point if you're interested in anything to do with hydrogen.
hydrogen.energy.gov

BP Solar
One of the world's leading photovoltaic manufacturers.
bp.com/modularhome.do?categoryId=4260

Sharp Electronics
The world's largest producer of photovoltaics.
solar.sharpusa.com

Real Goods Company
Solar Living Institute
Founded in 1978 this is one of the oldest and the best retailers of solar products and the first stop if you want to buy solar for your home.
gaiam.com/realgoods
solarliving.org

SunPower Corporation
One of the newest and best manufacturers of photovoltaics.
sunpowercorp.com

Spire Corporation
The world's leading builder of photovoltaic production machinery. Their factory is solar-powered.
spirecorp.com

General Electric Company
One of the world's primary producers of wind turbines, photovoltaic cells and water purification technologies.
ge.com
gewater.com
ge.com/products_services/energy.html

American Solar Energy Society
Solar Today is the top magazine of the solar industry.
ases.org

Hydrogenics Corporation
Manufacturer of fuel-cells and electrolyzers for hydrogen systems.
hydrogenics.com

United Technologies Corporation
UTC is a major corporation and the largest manufacturer of fuel-cells.
utcfuelcells.com

ISE Corporation
Supplier of hybrid-drive systems for buses and trucks.
isecorp.com

Autoblog Green
This site offers a wide range of articles and links to ongoing research in electric cars.
autobloggreen.com

Altairnano
A company involved in several technologies, most notably a new form of battery that can be charged quickly.
altairnano.com

Cobasys
A major manufacturer of new forms of batteries.
cobasys.com

Global Warming and Climate Change

Simmons & Company International
One of the leading investment banks in the energy industry. Matthew Simmons is
also extremely active in making public presentations about peak oil.
simmonsco-intl.com

Real Climate
Perhaps the best of the many Websites dealing with global warming.
realclimate.org

Pew Center on Climate Change
Some of the most elegantly clear writing on the subject of global warming.
pewclimate.org

American Council for an Energy-Efficient Economy
A Washington, DC-based group that lobbies for energy policy and technology.
aceee.org

Rocky Mountain Institute
Founded by Amory Lovins, one of the world's leading visionaries in energy issues.
Covers all modes of energy use, including transportation.
rmi.org

Environment and Ecological Restoration

Glen Canyon Institute
Focused on restoring the Colorado River.
glencanyon.org

California Water Impact Network
Nonprofits joining together to provide a town hall forum on water in California.
c-win.org

The Earth Institute at Columbia University
"Cross-cutting" interdisciplinary programs that meld environmental with
historical and technical studies.
earth.columbia.edu

TerraDaily
Eclectic but lively compendium of news and scientific items about the environment worldwide.
terradaily.com

Friends of the Los Angeles River (FoLAR)
Provides information on the river's current condition and status.
folar.org

Growing Native
The pathway to growing native plants in California. Basic information on native plants, the plant communities they live in and the experiences of those who plant them.
growingnative.com

The Rewilding Institute
An organization on the forefront of protecting wild species and expanding wilderness.
rewilding.org

Low Impact Development Center
Focus is on new strategies for dealing with water runoff, such as permeable paving and related technologies.
lid-stormwater.net

Ecological Restoration
The longest running and best magazine on-line concerning eco-restoration.
ecologicalrestoration.info

US Environmental Protection Agency, River Corridor and Wetland Restoration
An excellent guide to the basic principles of restoring aquatic ecosystems.
epa.gov/owow/wetlands/restore/principles.html

Western Rangelands Partnership
A group that's primarily focused on rangeland management in the western US.
rangelandswest.org

Loyola University, Environmental Studies/Sciences Program
This may be the first university-level program focused on themes of ecological restoration.
luc.edu/envsci

The Land Institute
One of the best sites on ecological restoration and one of a few focused on the science of prairies.
landinstitute.org

Louisiana Coastal Area, Louisiana Ecosystem Restoration
A government site on the restoration of Louisiana's coastline.
lca.gov

Chesapeake Bay Program
Chesapeake Bay is perhaps the biggest and most complex restoration project in the US.
chesapeakebay.net

San Francisco Bay Joint Venture
A joint restoration effort of 27 organizations in the San Francisco Bay Area.
sfbayjv.org

Transportation

General Motors Corporation
This site includes some explanation of new fuel-cell cars, including the Sequel.
gm.com
gm.com/company/onlygm/energy/cell.html

California Fuel Cell Partnership
The leading organization in the world developing hydrogen fuel-cell vehicle infrastructure.
cafcp.org

US Department of Transportation, Bureau of Transportation Statistics
The primary source for an overview of what moves where, when and by what means.
bts.gov

Railway Age
One of two trade magazines on railroads and related activities in public transportation.
railwayage.com

Texas Transportation Institute, Texas A&M University
The best of the handful of universities that include transportation departments.
tti.tamu.edu

Boeing
Builds very efficient airliners, such as the 787 Dreamliner, and developing a fuel-cell electric airplane pictured at the second link.
boeing.com
boeing.com/news/releases/2007/q1/070327e_nr.html

Association of American Railroads
This site provides information on major freight railroads in the US.
aar.org

American Short Line Railroad Association
Industry association representing over 400 small railroads in the US. Many innovations are occurring on these smaller railroads.
aslrra.org

Media
Treehugger
Extensive resources on all sorts of environmental strategies, e.g., a Sanyo dishwasher that uses no water and a Japanese condominium complex that is all solar-powered.
treehugger.com
treehugger.com/files/2006/02/sanyos_aqua_was.php
treehugger.com/files/2007/06/japans_first_so.php

Physorg.com
An excellent site for breaking news in physics, earth sciences, space science and a wide range of technology.
physorg.com

Encyclopedia of Earth
A good source of news on general environmental issues.
eoearth.org
eoearth.org/article/Green_roofs

Biofuelwatch
This site is focused on the damage caused by biofuel production.
biofuelwatch.org.uk.

Culture Change
A long-running newsletter, now a Website, encompassing environmental, economical and technological issues.
culturechange.org

Unusual or Special Interest

Natural Logic
A consulting firm that focuses on business strategies for complying with and going
beyond environmental regulations to achieve improved performance, lower costs,
higher profits and far less waste of any kind.
natlogic.com

US Green Building Council
Leadership in Energy and Environmental Design (LEED) building standards.
usgbc.org

Eco-Structure
This magazine is a good source of information on all manner of green building
products.
eco-structure.com

Natural Resources Defense Council
A long established environmental group. Their Santa Monica, California, office
building is itself a demonstration of green architecture.
nrdc.org
nrdc.org/cities/building/smoffice/resources.asp#smtop

World Watch
A Washington, DC-based organization that researches a wide range of environmental, energy, agricultural, transport and manufacturing issues.
worldwatch.org

Index

group intelligence, 197
Growing Native, 210

H

harbor tides, as source of power, 111
Harper's magazine, 22
HaveBlue, California, 111–112
health care costs, reduced, 14, 29
health, of species, 153, 157–163, 183; *see also* extinction
heat, 15, 55; production of, 47, 65, 70, 84, 89, 175; solar, 51, 69, 89, 176
heat-island effect, reducing, 98
Heat Island Initiative, 39
heliostats, 88
hemp, as consuming carbon dioxide, 148
hemp oil, 140, 148
Herman Miller furniture company, 34
hexavalent chromium, 162
highways, 7, 114, 119, 135, 186; *see also* Interstate highways
Hindenburg, 68
homes; solar-powered, xiv, 67; using energy-water systems, 175; *see also* energy-water systems; photovoltaic cells
Honda Motors, 107, 176–177
Hudson River sewage treatment system, 82
humans, pollutants in them, 76
Hurricane Katrina, 137
hybrid crops, 141
hybrid-drive systems, for vehicles, 208
hydrocarbons, burning of, 68
hydroelectric power, 49, 65, 67, 79, 108, 148
hydrogen, 12, 68–69, 107, 178; generating, 86, 107, 109, 112, 125, 157, 180; pipelines for, 85, 88; as renewable energy source, 66–67, 71, 97, 110; for storing electricity, 31, 62, 67, 113, 146; in vehicle engines, 112, 172; as water component, 67–68, 84–85, 109, 162; *see also* hydrogen fuel cells; hydrogen power systems
hydrogen-electric vehicles, 106, 108, 151, 174, 177–178, 188
hydrogen-electric systems, 109
hydrogen fuel cells, 47, 66, 70, 105, 109, 172; for electric vehicles, 107, 110, 211; *see also* hydrogen; vehicles

hydrogen fueling stations, 107, 174
hydrogen power systems, 205, 208, 111; *see also* hydrogen, as renewable energy source
hydrogen storage systems, 69, 173
Hydrogenics Corporation, 208
Hyundai Group, 107, 176
Hywire concept car (GM), 106

I

ice caps, melting of, 13, 155
India, consuming oil, 11
indium, 54
industrial ecology, 123, 126, 196
infrastructure, xv, xvi, 5, 7, 12, 125, 174, 181, 184–186, 189, 192; as changing, 17, 171, 174; as commons, 7–8; for energy-water system, 180; for solar energy, 188; for vehicles, 103, 114, 121, 189; for water, 174, 188–189
infrastructure technologies, 83; decentralized, 190
insulation, 37–39, 43
intercity passenger railway, 115, 183; *see also* passenger train system; railways
Intergovernmental Panel on Climate Change, 15
Internet, *see* World Wide Web
Interstate highways, 17; *see also* highways
irrigation, 26, 90–91, 146
ISE Corporation, 208

J

jet planes, 8, 17

K

Korean War, 17
Kyocera Solar Inc., 50, 206

L

Lake Powell, 146
Lambda Energia, 207
Land Institute, The, 211
land use, 7, 9, 196
landscaping strategies, 43, 121, 142–143
lead, 162
Leaf, lamp, 34
LEED (Leadership in Energy and Environmental Design), 126, 213

Leopold, Aldo, 12
life, 9, 195; quality of, xiv, 14, 22–23, 31, 33, 43
light, 55, 121, 195; *see also* sunlight
light-emitting diodes, 35, 53
Lighting Research Center, Rensselaer Polytechnic Institute, 205
local grid, *see* grids
locomotive, fuel cell, 107; *see also* fuel cells; hydrogen fuel cells
long-distance grids, *see* grids
Los Angeles Department of Water and Power (LADWP), 41–42
Los Angeles River restoration, 149, 151
Louisiana Coastal Area; Louisiana Ecosystem Restoration, 211
Lovins, Amory, 209
Low Impact Development Center, 210
lowlands, flooding of, 144
Loyola University; Environmental Studies/Sciences Program, 210

M

Machu Picchu, 3
manufacturing, computer-aided, 129
Marshall Plan, 16
mass media industry, xv, 5, 9, 16, 156, 193, 198
Maycock, Paul, 206
mercury, 162
metal hydrides, 107, 112–113, 205
methane, 11–13, 15, 71, 84, 89, 92, 200
Miasole, 176
microbank movement, 191–192
micro-machines, for pollution clean-up, 164–165
Mir space station, 200
Mississippi River system restoration, 137, 139
Mitsubishi Motors, 50
Moscone Center, San Francisco, 63
Motech, 51
Mother Earth News, 183
motors, electric, efficiency of, 109
music, digital, 17

N

Nanosolar, 54, 176
National City Lines (NCL), 101–102

native plants, 210
natural gas, 58–59, 66–70, 143; as energy source, 49, 71, 108; producing hydrogen, 69, 107
natural gas fuel cell, 105
natural gas power plants, 57
Natural Logic, 213
Natural Resources Defense Council, 213
nature, control of, 196, 203
Nissan Motors, 107, 176
nitrogen, 11–12, 68, 71, 89, 134, 143
non-imaging concentrator, 65
nuclear energy, 25, 29, 39, 49, 57–59, 71, 108, 196–198
nuclear fuels, 55, 59, 72, 108
nuclear industry, as safe, 71
nuclear power plants, 23, 54, 71

O

ocean thermal gradient plants, 49
off the grid, 50; *see also* grids
oil, 25, 47, 49, 66, 143, 170, 188; consumption of, 11, 59, 71; dependence on, 11, 27, 141; peaking of, *see* peak oil; rising price of, 58, 117, 121, 173; supplies, declining, xiv, 11, 13, 25, 57–58, 153, 155, 185, 188
oil companies, 170, 177–178, 180
oil fields, 21–22
Orchid Hotel, Mumbai, 126
organic farming, xiv, xv, 14, 141–142, 158, 161, 176, 179, 183, 197, 205
organic permaculture, 142
Orient Express, 187
Ovonic batteries, 67
Ovonics Hydrogen Solutions LLC, 107
Owens Lake, 42
Owens River watershed, 41–42
oxygen, 12, 67–68, 84–85, 109, 162
ozone, 12

P

Pacific Gas and Electric Company, 171, 206
passenger train system, 98, 100, 116, 118–119, 187, 205; *see also* intercity passenger railway; railways
pavement, permeable, 85, 149–150, 210
peak oil, 11, 58, 104, 129, 195, 205, 209
Penn Station (Manhattan), 99

permaculture farming, 143, 157; *see also* farming

permafrost, melting, 11

permeable paving, 85, 149–150, 210

personal security, 161, 170

pesticides, 140–142, 161, 163

petrochemical plants, cleanup of, 140

petroleum, replacing, 136

Pew Center on Climate Change, 209

photovoltaic (PV) cells, 12, 14, 51–53, 69, 72, 110, 183, 190–191, 205–208; decentralized control of, 197; efficiency of, 60, 63–64; as generating electricity, 31, 53, 62, 67, 69, 83, 85, 87, 89, 111, 128; powering external appliances, 108, 112; as renewable energy, 49, 54–55, 65, 92, 178; on roofs, 106, 109, 120, 172; as roofing, coverings, windows, 35, 53–54, 70, 87, 111, 148, 171–173, 175; *see also* photovoltaic panels

Photovoltaic Energy Systems, 206

photovoltaic industry, 51, 126, 176

photovoltaic panels, 67, 87–88, 171

photovoltaic technology, 24, 66–67, 70–71, 86, 173–174

Physorg. com, 212

plastics, biodegradable, 156

pollutants, recycling of, 87

pollution, 21, 60, 71, 89, 98, 106, 197; clean-up using micro-machines, 164; ended by energy-water system, 93; *see also* air pollution; water pollution

population growth, endless, 14

power plants, 7, 25, 54, 61, 73, 80, 89; *see also* coal power plants; natural gas power plants; nuclear power plants

profits, 169–170; from renewable-energy products, 171, 180

propane, 170

proton exchange membranes, 205

public transportation, *see* transit system, public

PVs, *see* photovoltaic cells

Q

Q-Cells, 50

R

radioactive waste, 25, 71

radium, 162

railroads, *see* railways

Railway Age, 211

railways, xiii, xiv, 17, 97, 101–102, 123, 174; efficiency of, 99, 104, 115; expansion of, xv, 117–118, 135, 137, 186, 188; as infrastructure, 7–8, 114, 116, 121, 184, 211; revitalization of, xiv, 16, 98, 103, 117, 179, 187, 197; *see also* retail railway

rainforests, clearing of, 136

rainwater, 80, 91, 171, 195; captured, 85, 150, 175; generating electricity, 125

rainwater catch basins, 42, 79

rangeland management, 210

Real Climate, 209

Real Goods Company, Solar Living Institute, 207

recycling, as practice, 197

refrigerators, as energy consumer, 38–39; *see also* appliances

regenerative braking, for car, 109

religion, as tie, 194

renewable energy technologies, 31, 54, 57, 63, 69, 70–71, 177; *see also* energy, renewable; energy technologies

resins, biodegradable, 140; for plastic roofing, 148

resource consumption, 11, 14, 22, 118, 170

restoration, *see* ecological restoration

retail railway, 113, 120–121, 183; *see also* railways

Rewilding Institute, The, 210

riverside environments, restoring, 138

Rocky Mountain Institute, 209

roofing; impermeable, 78–79; of photovoltaic material, 72–73, 88, 148, 173

roofs, 30, 60; with PV cells, 63–64, 70, 73, 83, 85, 125, 171–172, 190; for rainwater capture, 82, 190

roofs, sawtooth, 125

S

salmon, 148

saltwater intrusion, by dams, 139; *see also* dams; flooding

San Francisco Bay Joint Venture, 211

Sandia Labs, 107

Sanyo Electric Co. Ltd., 50, 206, 212

Sanyo streetlight, 35

satellites, 7, 17

savannahs, clearing of, 136

uranium, 59, 66, 162, 170

US Army, and fuel cell, 107

US Army Corp of Engineers, 137

US Department of Agriculture, 140

US Department of Defense, 13

US Department of Energy, 205, 207

US Department of Justice, 102

US Department of Transportation, 116; Bureau of Transportation Statistics, 211

US Environmental Protection Agency, 163; River Corridor and Wetland Restoration, 210

US Green Building Council, 213

US National Renewable Energy Laboratory, 55

US Navy, 198

utility companies, 49, 57, 64, 66, 92, 170, 176–178, 180; *see also* grids

V

vapor barriers, 38

vehicles, 105; electric, 14, 107, 151; ethanol-burning, 136; using hydrogen-fuel cells, 66, 107, 110, 151, 174, 211; *see also* cars

Verne, Jules, 198

VIA Rail, 102

Virgin Rail, 98

vision for future, 23, 19, 183–184, 197, 191, 194, 199

Volkswagen AG, 107

W

walls, impermeable, 79

wastewater, removal of, 77; *see also* water treatment systems

water, xv, 180, 188, 196; bottled, 7, 28, 80, 88, 90; conservation of, 40, 42, 92, 176; for drinking, 87, 90, 111, 163, 173; as energy source, 109, 146; as life element, 3–4, 6, 8, 19, 77, 194; as producing hydrogen, 112, 173; purification of, 31, 14, 71, 176, 208; recycled, 108, 110, 176, 190; shortage of, 21, 23, 31, 78; *see also* water filtration technology; water treatment systems

water filtration technology, 173, 176, 178

water pollution, 8, 75–76, 78, 87, 162, 170

water, pure, 162, 178; from energy-water system, 22, 84, 175; from fuel cells, 67, 70, 84–85, 125, 171; as piped in, 28, 78–79; as plentiful, xv, 21, 26, 30, 89

water technology, 9, 24, 31, 93, 170, 173, 175; financial profit in, xiii, xv, 171, 177

water treatment systems, 40, 77, 82; *see also* wastewater

water turbines, 86

water vapor, 11, 76, 109, 113; contributing to climate change, 108; from hydrocarbons, 68, 89, 112

weather, 15, 48, 70, 80

Web, *see* World Wide Web

Western Rangelands Partnership, 210

wetlands, 134, 143–144; restoration of, 137–139, 148, 176, 192

wild agriculture, 142

wild ecosystems, xv, 142, 147, 151

wild fisheries, 134, 148

wild game ranching, 14

wild grasslands, 147

wild protein, 149

wild rivers, 151

wild species, protecting, 210

wind, as commodity, 171

wind power technology, 54, 65, 67, 88–89

wind turbines, 23, 49, 52, 62, 66, 69–70, 86, 88, 92, 125, 175–176, 178, 208

windmill farms, 72, 174, 183

windows, multipane, 37–38

wood fuel, 68

World War II, 16–17

World Watch, 213

World Wide Web, 16–17, 25, 126, 159, 176–178, 183; as knowledge repository, 128–129, 156, 189

Wyoming grasslands, restoring, 10

X

xeriscape strategies, 40

Y

Yosemite Valley restoration project, xiii

YV 88, xiii

Z

zinc, 162

About the Author

CHRISTOPHER C. SWAN resides in San Francisco and was born in Berkeley. He has written, illustrated and produced three books — *Cable Car* (Ten Speed Press '73), *YV88* and *Suncell* (Sierra Club Books, 1977, 1987) — and several dozen articles on ecological restoration, energy, transportation and water issues in national and California publications. As a resident of Northern California, Christopher has long focused on the realities of the state, from the Bay Area to the Sierra Nevada foothills, where he lived with his wife and two sons. As an entrepreneur focused on railway development, Chistopher founded Suntrain in 1980 and proposed several railway projects, notably a 1991 plan for a railway serving Yosemite, with track carpeted in grass and trains powered by solar energy, and all designed to allow restoration of Yosemite Valley. In 2007, Christopher is focused on freelance design projects and new railway ventures.

If you have enjoyed *Electric Water*, you might also enjoy other

BOOKS TO BUILD A NEW SOCIETY

Our books provide positive solutions for people who want to make a difference. We specialize in:

Sustainable Living • Ecological Design and Planning
Natural Building & Appropriate Technology
Environment and Justice • Conscientious Commerce
Progressive Leadership • Resistance and Community • Nonviolence
Educational and Parenting Resources

New Society Publishers

ENVIRONMENTAL BENEFITS STATEMENT

New Society Publishers has chosen to produce this book on recycled paper made with 100% post consumer waste, processed chlorine free, and old growth free.

For every 5,000 books printed, New Society saves the following resources:[1]

33	Trees
2,971	Pounds of Solid Waste
3,269	Gallons of Water
4,264	Kilowatt Hours of Electricity
5,401	Pounds of Greenhouse Gases
23	Pounds of HAPs, VOCs, and AOX Combined
8	Cubic Yards of Landfill Space

[1]Environmental benefits are calculated based on research done by the Environmental Defense Fund and other members of the Paper Task Force who study the environmental impacts of the paper industry.

For a full list of NSP's titles, please call 1-800-567-6772 or check out our web site at:

www.newsociety.com

NEW SOCIETY PUBLISHERS